HOLLEY® 5200 SERIES HANDBOOK

including 5210, 5220, 6500, 6510 and 6520

CONTENTS

Published by HPBooks, P.O. Box 5367, Tucson, AZ 85703 602/888-2150
ISBN: 0-89586-050-3 Library of Congress Catalog Card No. 81-82587

Editorial Director: Tom Monroe; Editor: Jim Miller; Art Director: Don Burton; Book Design: Tom Jakeway; Book Assembly: Ken Heiden; Typography: Cindy Coatsworth, Michelle Claridge; Photos: Mike Urich, Bill Fisher, Howard Fisher, Jim Miller. Cover Photo: Jim Miller.

INTRODUCTION

Typical Model 5200 used on in-line-4 and V6 Ford engines. Wiring harness is for electric assist for the hot-water choke.

This Italian became a naturalized U.S. citizen in 1970, when Holley obtained a license from Weber to build the Model 5200 for the 2.0-liter German Ford engine used in the Pinto. That was the beginning, and today the 5200 and related models are among the most popular carburetors in original-equipment use and have become highly "Americanized."

After that first Pinto, Ford also used the 5200 on the V6 and 4-cylinder Capri, Mustang II and Cortina.

Those were all 5200s. The 5210 was introduced in 1973 on the L-11 performance version of the 2.3-liter Vega. In 1977, the 5210 was used on the new Pontiac 2.5-liter "Iron Duke" and the 2.0-liter Audi four used in the AMC Gremlin. In 1978, the 5220—same basic family—was introduced on the 1.7-liter VW/Audi engine in the Dodge Omni and Plymouth Horizon.

Holley makes aftermarket versions for most of these applications, including some special economy versions. Holley has also introduced a line of replacement carburetors for a number of Japanese passenger cars and trucks.

Clever enthusiasts and a number of aftermarket suppliers have adapted the 5200 to other engines, particularly air- and water-cooled VWs.

I deal with several models of the 5200 here, but there is little difference between them. So, throughout this book I'll be treating them pretty much as if they were one model—the 5200—freely mixing models in all sections. I'll note when there are significant model differences.

I'll give you some history of the carburetor, describe the basic systems, feature a few applications and guide you through a teardown, repair and adjustment. I'll also give you a little basic theory and a few tips along the way.

This is a rather complicated little carburetor with a great number of parts. But complication can be turned into simplicity if you take it one step at a time.

Airflow Capacity—The 5200, although small in capacity on an absolute basis, is relatively large for the small-displacement engines using it. I'm not the least bit bashful about referring to the carburetor as "high performance."

On most models primary-venturi diameter is 26mm (1.02 in.) and the secondary is 27mm (1.06 in.). Throttle-bore diameters are 32mm (1.25 in.) and 36mm (1.4 in.) with a 43mm (1.7 in.) center spacing on a metric flange. The stud spacing measures 47mm x 93mm (1.84 in. x 3.66 in.).

When introduced, this carburetor came in only one size—270 CFM at three inches of mercury. In 1978 a 235-CFM model was

Original Model 5210 for Vega 2.3-liter engine. Typical GM integral fuel filter is located at the front. Like the Ford carburetor, this one has the hot-water choke mounted at right.

introduced on Ford and Mercury 4-cylinder engines, with a 23mm primary venturi and 27mm secondary. Since 1979, Chevettes have been equipped with a 245-CFM model, with primary and secondary venturis of 23mm and 29mm.

Airflow Rating—Before I go much further I should clear up the mystery about airflow ratings. Flow rating is a much more effective sizing method than venturi diameter because many things in the air section—nozzles, boost venturis and venturi angles—affect flow capacity.

The generally accepted method of rating carburetor size or capacity

This Model 5210 is used on 1977 and later Pontiac Astre and Sunbird 4-cylinder engine. Note the all-electric choke 1, and solenoid-assisted bowl vent. Tube 2 hooks to evaporative canister.

This "octopus" is the Model 5220 used on the Dodge Omni and Plymouth Horizon. All-electric choke is at right. Black cylinder below is a throttle-position sensor for electronic ignition.

Additional wiring is for air-conditioning controls. When A/C is on, throttle position is increased at idle. A/C is shut off near wide-open throttle to improve power.

Basic two-piece construction. Throttle body is integral with the main body (left); air horn with choke plates, float and fuel inlet is at right.

is by flow rate *at a given pressure drop.* The carburetor is mounted on a test stand with the throttles held wide open. Airflow through the carburetor is adjusted to a specific *pressure drop,* or difference in pressure, between the base of the carburetor and the atmosphere. The number of cubic feet per minute—CFM—of air passing through the carburetor under these conditions is its airflow rating.

Holley carburetors are rated in CFM, but the rate is really *mass flow* because it is CFM at a specific pressure and temperature.

One- and two-barrel carburetors are rated at a pressure drop of 3 inches of mercury (''Hg) or 1.5 psi. Four-barrel carburetors are rated at 1.5''Hg, or 0.75 psi. To determine what a carburetor rated at 1.5''Hg flows at 3''Hg, simply multiply the flow rating by 1.4. To find what a carburetor rated at 3''Hg flows at 1.5''Hg, divide the rating by 1.4.

Remember, to make a valid comparison of components, knowing the flow rate is not enough. You must know the pressure drop across the component when the flow rate was measured. At relatively low pressure drops, you can assume the *airflow changes directly as the square root of the pressure drop across the component.*

I bring this up because people are always trying to compare two- and four-barrel carburetors without making the conversion. For

Bottom view showing primary throttle on the left, larger secondary on the right. This is a metric-standard flange with both bore and stud spacings different from any SAE standard.

5200 includes booster venturis 1; accelerator-pump shooters (only primary is drilled but secondary can be opened up) 2; main air bleeds 3; main-well tubes 4, fit under air bleeds.

This particular 5200 has primary and secondary idle systems, with idle air bleeds 5; removable idle jets, 6; in individual holders, 7. Secondary-system power well 8; deceleration-system well, 9. Power valve, 10, and main jets, 11, screw into bottom of bowl.

example, a 600-CFM four-barrel has more flow capacity than a 600-CFM two-barrel—40% more—due to the different pressure ratings.

This same flow-rating method is used for other parts of the induction system besides the carburetor, such as air cleaner, intake manifold and cylinder head.

Construction—Basic construction of the 5200 carburetor is two zinc castings, the main body—including the throttle body—and a separate air horn.

These staged two-barrels have a mechanically operated secondary throttle, which I'll explain later when I talk about secondary systems. Primary and secondary venturis are different sizes, as is common for progressive-secondary carburetors, both two- and four-barrel.

A diaphragm-type accelerator pump is used. I'll feature this item in its own section. Connections are provided for coolant or electricity to the automatic choke, fuel inlet and return, timed and vacuum spark, and for a PCV valve. A connection for the deceleration valve used on some Ford products is also provided.

The air-cleaner mounting flange is unusual if you're used to thinking in circles—it's not round. If you've been servicing US-designed carburetors, the screw-in main air bleeds and idle-feed restrictions may also seem unusual.

Most carburetors are identified here, though some use a tag as shown at left.

The "R8263" (right photo) is an engineering, or list number. Holley Engineering releases a parts list to the production plants. The carburetor is built to this list which contains everything right down to the smallest detail and specification. Hence the name "list."

Some normal replacement carburetors are cataloged and packaged under different "sales" numbers, but the engineering list or part number is always the one stamped on the carburetor. Specification information given in the various Holley literature also refers to the list number.

The last four digits are the manufacturing date. The first three are the day of the year. 046 means the 46th day or Feb. 15. The last number is the year (9 is for 1979).

Center digit, 2, indicates a slight change from the basic list number and is not always used.

Most carburetors are identified at this location, or on a tag as shown at left.

1 FUEL-INLET SYSTEM

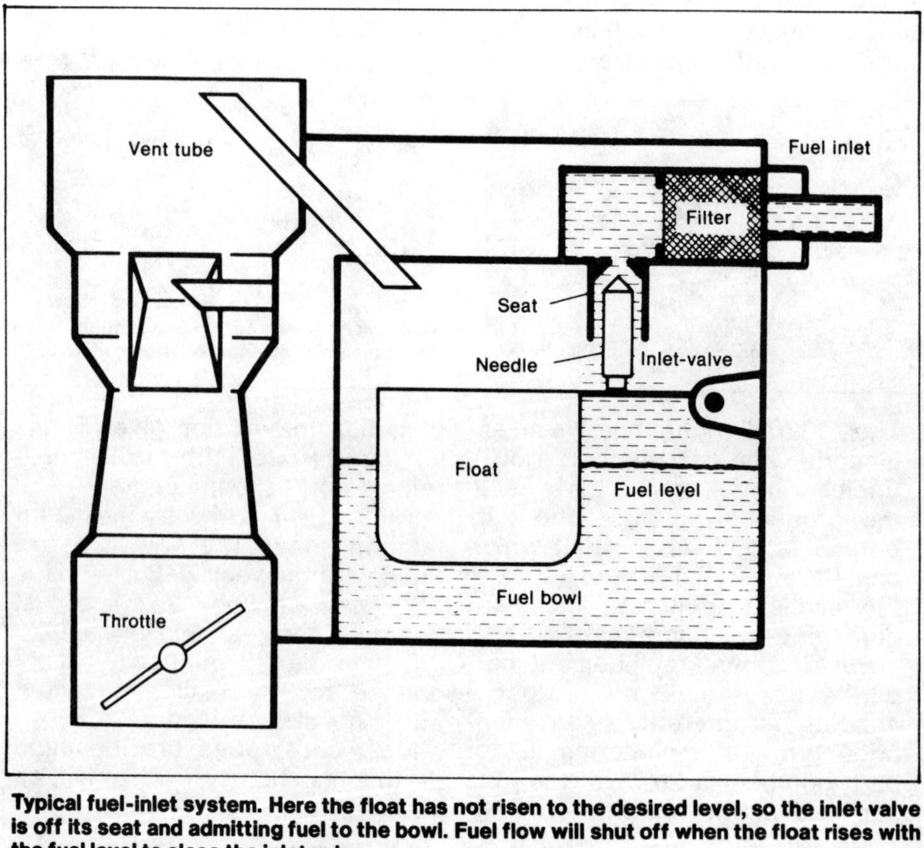

Typical fuel-inlet system. Here the float has not risen to the desired level, so the inlet valve is off its seat and admitting fuel to the bowl. Fuel flow will shut off when the float rises with the fuel level to close the inlet valve.

The fuel-inlet system includes the fuel bowl and float, the fuel-inlet fitting and the inlet-valve-and-seat assembly. These are the basic components, but other features are added in some cases.

Fuel-bowl Venting—Fuel bowls are vented to the air horn of the carburetor to maintain a constant "reference" pressure. This ensures the fuel/air ratio remains fairly constant as the air filter gets dirtier. This doesn't mean a dirty filter isn't harmful—it still restricts inlet airflow. This restriction causes a pressure drop, and consequently a less-dense charge and loss of power.

In some cases the bowl is also vented to an outside canister at idle and engine-off conditions. Here, fuel vapors are stored as a means of controlling evaporative emissions. During normal operation vapors are drawn back into the engine through the carburetor *purge port* and burned.

Some versions have a fuel-return connection just upstream of the fuel inlet. A portion of the fuel supplied by the pump is returned to the tank to prevent vapor lock and ensure liquid fuel is supplied to the carburetor.

This fuel return is usually used on air-conditioned cars due to their higher under-hood temperatures. Vapor forming in the line between the fuel pump and carburetor during a hot engine-off condition is returned to the tank.

Inlet-System Operation—Fuel passes through a screen or filter and enters the fuel bowl through the inlet valve. The 5210 uses the standard GM-type internal

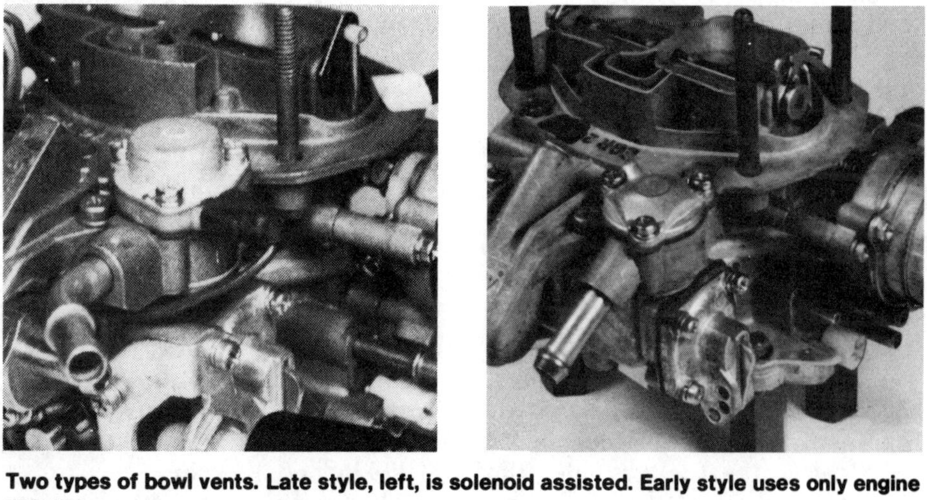

Two types of bowl vents. Late style, left, is solenoid assisted. Early style uses only engine vacuum.

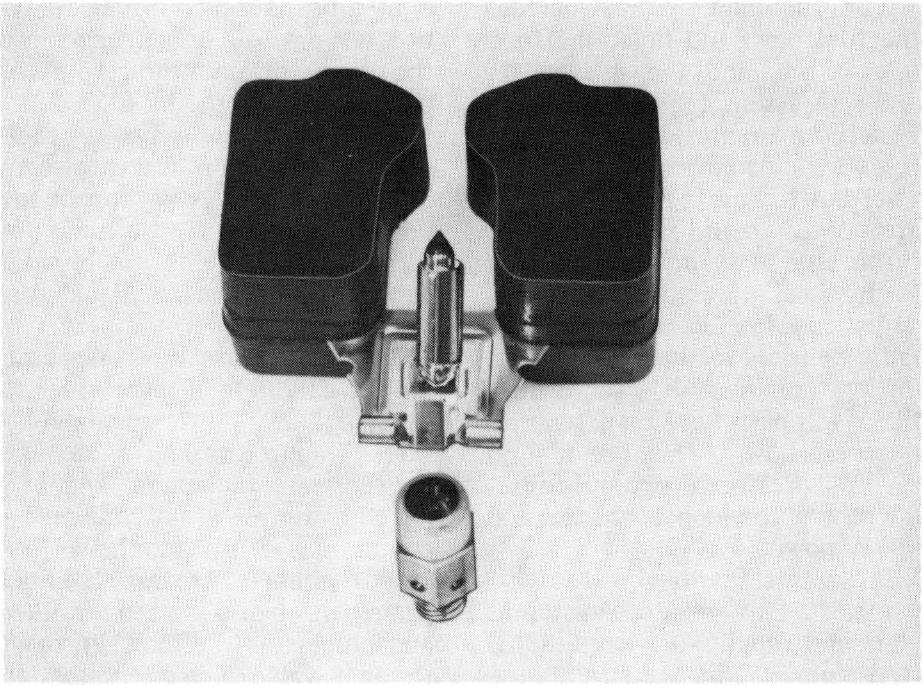

Fuel-inlet seat shown at bottom. Above is the "two-lung" float assembly with the inlet valve attached by a clip.

sintered-metal or paper filter, but other 5200 models use a screw-in external filter.

The fuel bowl is a reservoir. As fuel enters the bowl, the float rises and closes the inlet valve when the level of fuel reaches a specific height. This level must be maintained for proper carburetor operation, but varies slightly from carburetor to carburetor.

Referring to the schematic, think of the float and inlet valve as a scale balance with the hinge pin as the pivot. Fuel pressure against the inlet valve provides the opening force while the buoyancy of the float in the fuel provides the closing force. Buoyancy force increases as fuel rises on the float. Fuel level is established by the balance of opening and closing forces.

Change this balance and you change the fuel level. For instance, increasing the inlet-seat size or inlet pressure raises fuel level, so the float setting should be lowered to compensate. A rule of thumb is that a 1-psi (pound per square inch) fuel-pressure increase gives a 0.03-in. increase in fuel level.

Inlet Valves and Floats—The inlet valve is spring-loaded to resist the vibration of 4-cylinder engines. The tip is Viton for better sealing. Three seat sizes are currently used on 5200s: 0.0785 in., 0.089 in. and 0.095 in.

The only one currently available in the Holley Performance Catalog

Inlet System

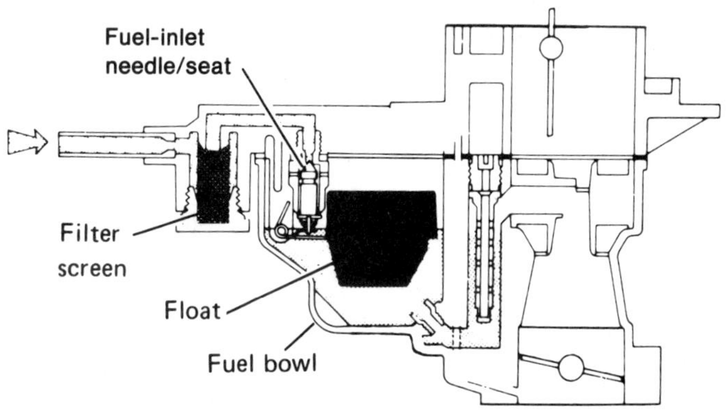

is the 0.0785-in. (2.0mm) seat sold as Part No. 18BP-256AS.

Increasing inlet-seat size is a good-news/bad-news story. The good news—fuel-flow rate will increase. The bad news—fuel-handling control will be sacrificed. Don't install a larger seat unless your engine really needs the extra flow. If you install a larger seat and get no measurable performance improvement, reinstall the original seat.

The float is made of a synthetic closed-cell material and has two "lungs." A small retaining clip, hooked over the end of the float-lever tang, is attached to the fuel-inlet needle. This ensures that the needle will unseat as the fuel level and float drop. When the float is down, a tang on the float lever contacts a horizontal bumper to stabilize the float against undue bouncing.

2 IDLE SYSTEM

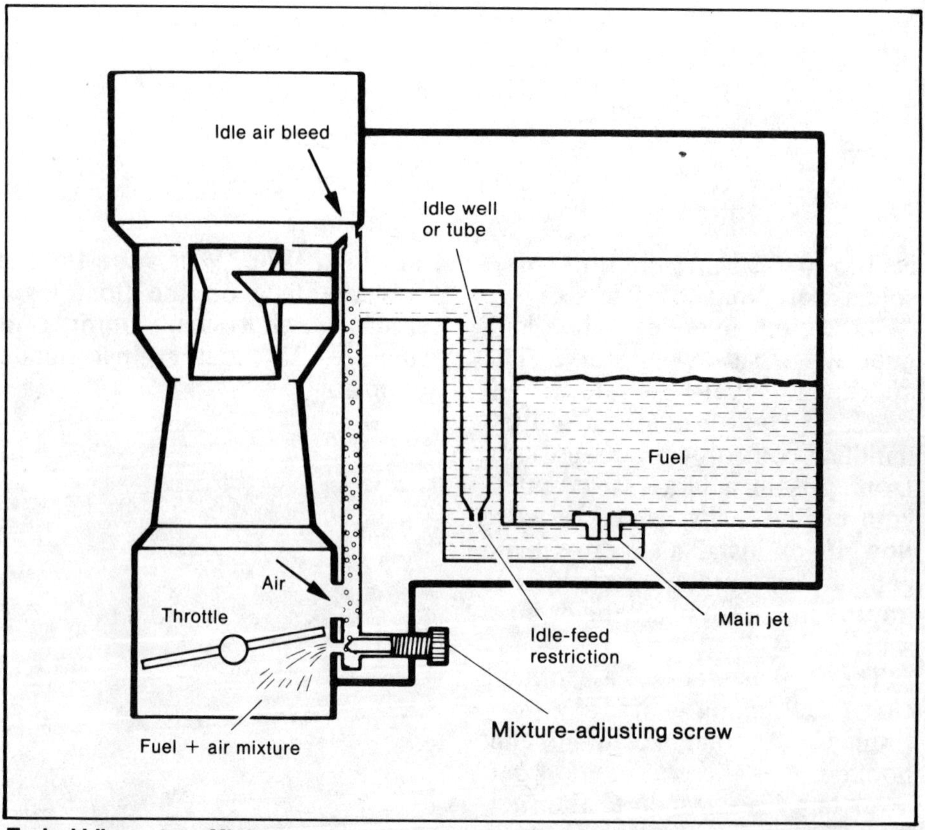

Typical idle system. Mixture-adjusting screw controls fuel flow at idle. Air bleeds are in the main body, open to the air horn. Although 5200 idle system is arranged differently from that shown, operation is identical.

The idle system serves several purposes. It provides a richer mixture at idle and supplies fuel in the transition from idle to part throttle. It also provides a small portion of the fuel under cruise conditions. Finally, it allows idle-mixture adjustment for engine tuning.

Some 5200s have an idle system on both the primary and secondary sides. Construction and operation are very similiar except the secondary system is non-adjustable, so the following description of operation applies to both.

Fuel flows from the fuel bowl through the main jets into the main well. From the main well a passage connects to the idle well. The fuel is metered as it passes

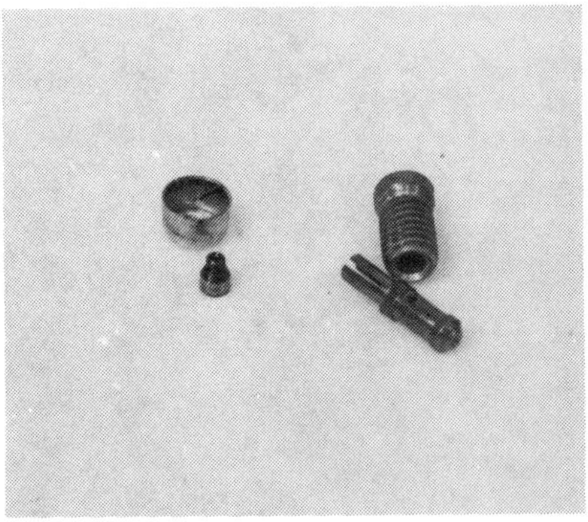

Most 5200s use either screw-in idle restriction (right) or pressed-in restriction with cup plug.

Idle passage (arrow) drilled and tapped to accept screw-in idle restriction.

Compare idle passage to that at left. Restriction presses into the smaller hole, while the larger, concentric hole accepts the cup plug.

Pontiac version shows a third type of idle restriction, at the end of a pressed-in tube under a lead plug (arrow). Corresponding passage on secondary side is open because this particular carburetor has no secondary idle system.

through the idle jet, located in a horizontal passage at the top of the idle well.

Just as the idle fuel starts down another vertical passage it is joined with air from the idle *air bleed*, a pressed-in restriction. Air bleeds *emulsify* fuel—mix air with fuel—to prepare it for atomization. This also lowers the effective viscosity of the fuel to encourage earlier feeding of the system.

The fuel/air emulsion then moves down the passage where some of it passes out into the throttle bore through transfer holes or slots when the throttle plate is open far enough to expose these to vacuum.

The remainder of the emulsion flows out the curb-idle discharge port. On the primary side, flow is adjusted by a tapered screw that extends into the port. The secondary side sometimes has transfer holes or slots, but never has a mixture adjustment.

Certain versions use screw-in idle jets, which are easily replaced. Others use pressed-in jets—a little more difficult to replace and not readily available. Screw-in idle jets are available in the Holley Performance Parts Catalog. Either way,

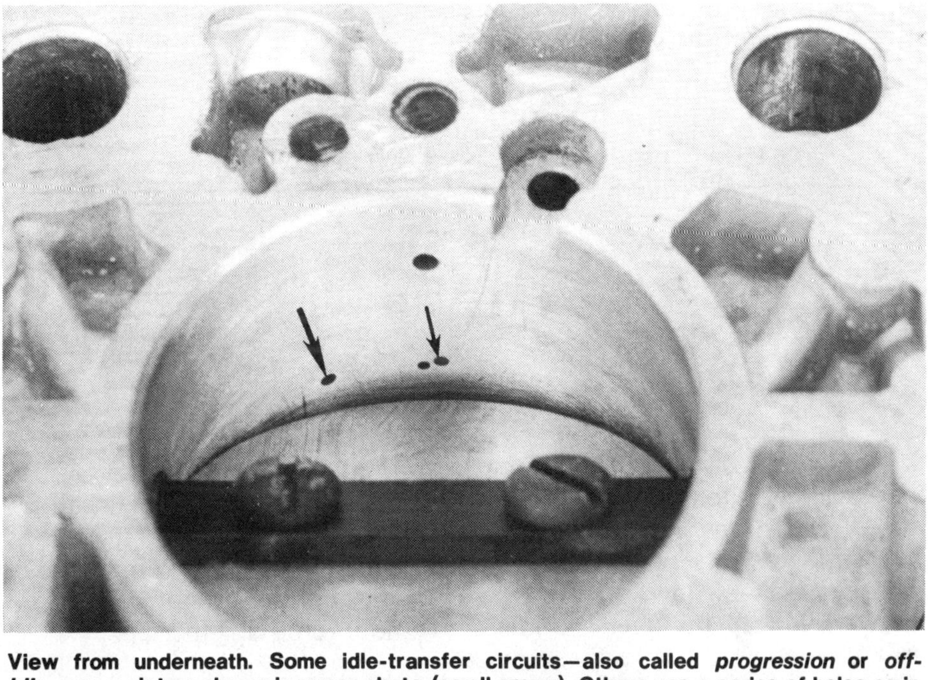

View from underneath. Some idle-transfer circuits—also called *progression* or *off-idle*—use a slot as shown in upper photo (small arrow). Others use a series of holes as in lower photo (small arrow). Curb idle is always the hole nearest the mounting flange. Large arrow in each photo indicates spark—timed vacuum—port.

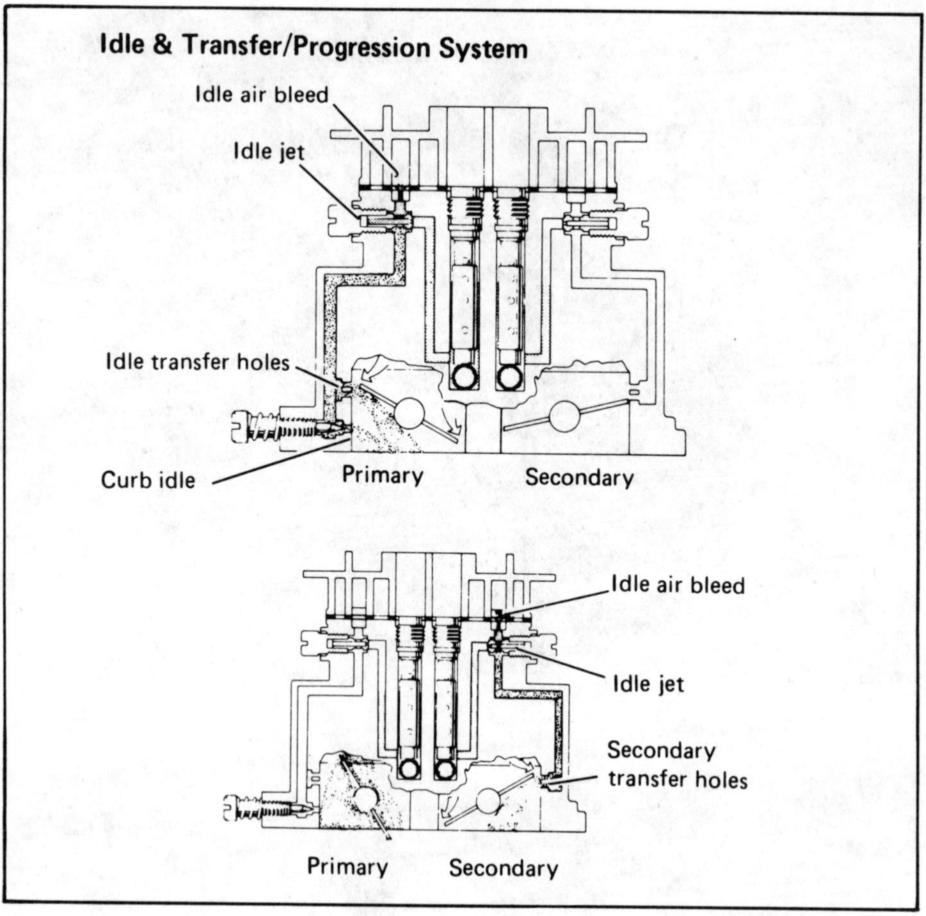

in most cases it shouldn't be necessary to alter the idle feed or bleed restrictions. But if you have to, work on the *feed*—fuel flow—not the *bleed*—airflow.

A plastic cap fits over the idle-mixture screw to limit adjustment to less than a full turn. This cap is called an *idle limiter* and it maintains idle-mixture adjustments to a very narrow band to help control emissions. On some applications the cap allows no movement at all, so idle mixture cannot be adjusted without removing the cap.

Idle-limiter cap. Tang on cap allows less than one turn of adjustment.

Some later models use this type limiter cap, which allows no adjustment. Idle is set during flow testing—a final operation on the production line.

3 MAIN METERING SYSTEM

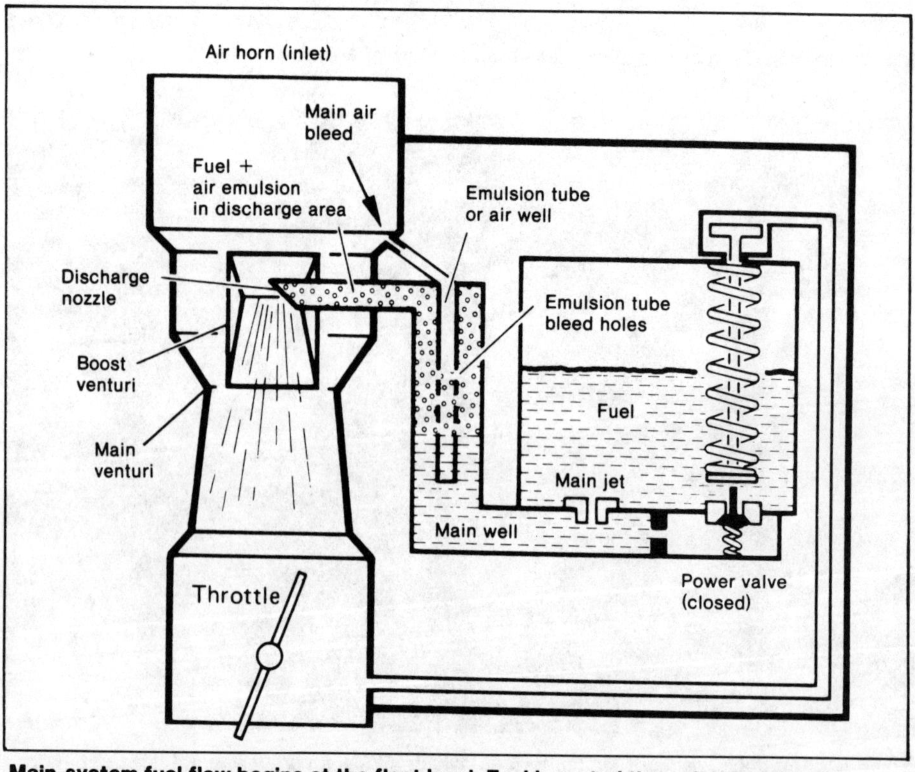

Main-system fuel flow begins at the float bowl. Fuel is routed through the main jet into the main well where the fuel is premixed with air. This fuel/air mixture is then routed to the booster venturi via the discharge nozzle. Although 5200 main metering system is arranged differently from that shown, operation is identical.

As the throttle plate is opened past the idle position, airflow increases and the main system begins to flow. Referring to the schematic, fuel flows from the fuel bowl, through the main jet into the bottom of the main well.

Here, the fuel is mixed with filtered air that enters the main-well tube through the main air bleed. The mixture of fuel and air moves up the main well and across and out the discharge nozzle located in the *boost venturi.*

I'll talk about the components of the main system one at a time.

Main Jets—These screw-in metering orifices control fuel flow into the main system.

There is a misconception among some people about jets—that size alone determines flow capacity. This is not the case. Although size is the major factor, surface finish and the shape of the jet entry and exit also affect flow. Holley checks each jet on flow-test equipment and rates it according to flow. A number is stamped on the jet to indicate this flow.

During the first few years this carburetor was built in the U.S. the jets were flow checked, but the stamped number represented nominal diameter in hundredths of a millimeter. These jets were cataloged with a 22R-103-XXX number, where XXX represented nominal diameter.

For instance, a 22R-103-130 jet —stamped **130**—has a 1.3mm nominal diameter. There is an actual *flow range* for this and all other jet sizes. These older jets are brass-colored.

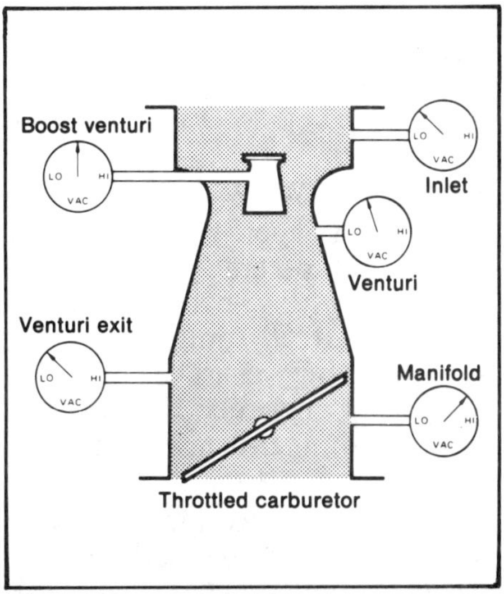

Typical vacuum drops inside a carburetor. Slight vacuum at inlet represents drop across the air cleaner. Gage at large venturi throat shows higher vacuum because air is still at relatively high velocity. Vacuum returns almost to the inlet value just before the throttle plate.

Throttled carburetor (above) shows very high manifold vacuum because a large pressure drop occurs across the partially opened throttle. Wide-open carburetor's low manifold vacuum (below) indicates a heavy-load/dense-charge condition.

In both cases, the highest carburetor vacuum is at the boost-venturi throat. The signal to the main system is picked up here.

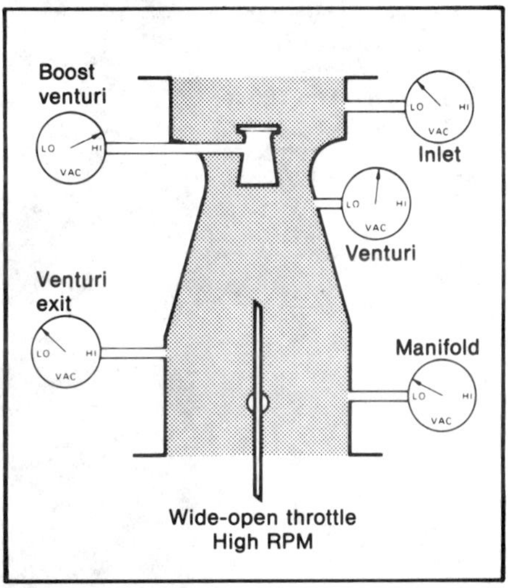

In 1975 Holley began using a series of main jets sized and marked according to *average flow.* They are dyed green to avoid confusion with the earlier brass-colored jets. A 124-325 —marked **325**—flows approximately 325cc (cubic centimeters) per minute of test fluid with a constant *head*—pressure—of 50 centimeters. The accompanying chart shows the relationship between flow-marked and the nominal-diameter-marked jets.

Drilling jets to increase their flow is not recommended. This destroys the entry and exit features to a certain degree. It may also introduce a swirl pattern even if the drill is held in a pin vise and turned by hand. Consequently, you cannot be sure of the flow of a modified jet unless you check it on a flow-test stand.

Two kinds of 5200 main jets. Brass-colored jets (left) used through 1975 are marked for diameter. Those on right are marked for flow and dyed green.

Main Air Bleeds and Well Tubes—These carburetors are equipped with screw-in dimensioned air bleeds. Just as with early main jets, they are stamped with numbers representing diameter in hundredths of a millimeter. Typical number is 123-XX. If the XX number is 45, as an example, the size is 0.45mm. A limited number of these bleeds are availa-

Removable threaded restrictions in the center are high-speed, or main, air bleeds. Stamped number is diameter in 1/100 millimeter. Primary air bleed marked 195 is 1.95mm in diameter, while secondary, marked 175 is 1.75mm.

Main air bleed with well tube. Don't modify the holes in the tube.

ble in the Holley Performance Catalog.

Because air bleeds introduce air into the fuel, changing them has the opposite effect of jet-size changes. Larger air bleeds lean the mixture and smaller ones richen it.

Air bleeds also affect the nozzle *spillover,* or *starting point*—the engine speed or rpm the nozzle *starts* to flow fuel. A larger bleed weakens the signal and delays the starting point to a higher rpm. For this reason, I recommend the tuner not use bleeds to richen or lean fuel/air mixture. For main-system recalibration, main jets are the only way—that's why they're changeable.

Just beneath the main air bleed is a removable main-well, or emulsion, tube with a number of holes in the side to introduce air into the main well. Don't alter the size of these holes. Their size has been arrived at through extensive testing and they are not metering orifices per se.

Venturis, Boost Venturis and Nozzles—The venturi, or throat, is the controlling sensor for the main system. It also determines carburetor-airflow capacity to a large degree.

Green Main Jets	Brass-Color Main Jets	High-Speed Air Bleeds	Idle Jets
124-227	22R-103-130	123-150	123-45
124-239	22R-103-132	123-160	123-50
124-243	22R-103-133	123-170	123-55
124-255	22R-103-135	123-175	123-60
124-263	22R-103-137	123-180	123-65
124-275	22R-103-140	123-185	123-70
124-283	22R-103-142	123-190	123-80
124-299	22R-103-145		
124-311	22R-103-147		
124-325	22R-103-150		
124-346	22R-103-155		
124-357	22R-103-157		
124-374	22R-103-160		
124-404	22R-103-165		
124-423	22R-103-170		
124-455	22R-103-175		
124-477	22R-103-180		
124-524	22R-103-185		

Brass-color jets use old numbering system and metric sizing:
130 = 1.3mm

Green jets listed in first column are the same flow capacity as the early brass jets shown in second column. For example, a 124-227 has the same flow rating as a 22R-103-130.

Main System

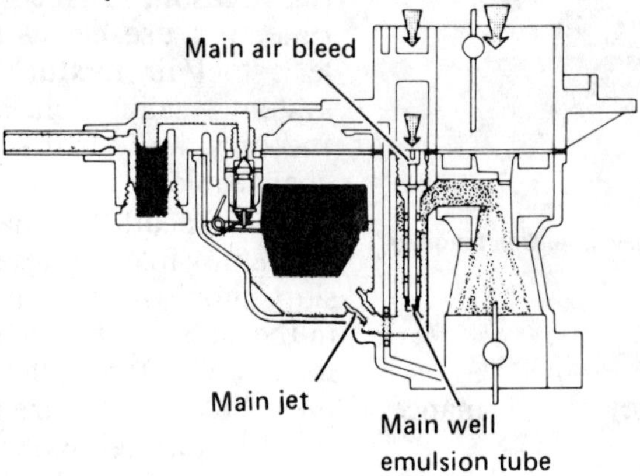

As velocity increases through the venturi, absolute pressure decreases—vacuum increases if you prefer. When the pressure drops low enough to lift the fuel to the spillover point, main-system flow begins. The greater the velocity, the greater the pressure drop and fuel flow.

A small venturi provides high velocities, giving strong metering signals and good atomization and vaporization. It provides good low-speed throttle response, but is more restrictive and reduces an engine's power potential.

A large venturi gives reduced metering signals and less effective atomization at low engine speeds, but is less restrictive and increases an engine's power potential. That's why carburetor size is a compromise. A little later I'll explain how this carburetor gives you the best of both worlds.

The boost venturi is a *signal amplifier* with the fuel-discharge nozzle located inside. As a venturi within a venturi, the boost venturi "sees" a lower absolute pressure than the main venturi. To increase this effect, the boost venturi's discharge is located at the point of lowest pressure in the main venturi. The placement of the boost venturi in the throat of the main venturi is extremely important and should not be changed.

Some 5200 carburetors have a double boost venturi. The purpose here is further amplification of the metering signal. More about this in the discussion of Economaster carburetors.

Pressed-in booster—this is the primary side. Relationship between the bottom of the booster and main venturi throat is extremely important.

Double boost venturi gives a stronger primary metering signal, but sacrifices some airflow.

4 POWER SYSTEM

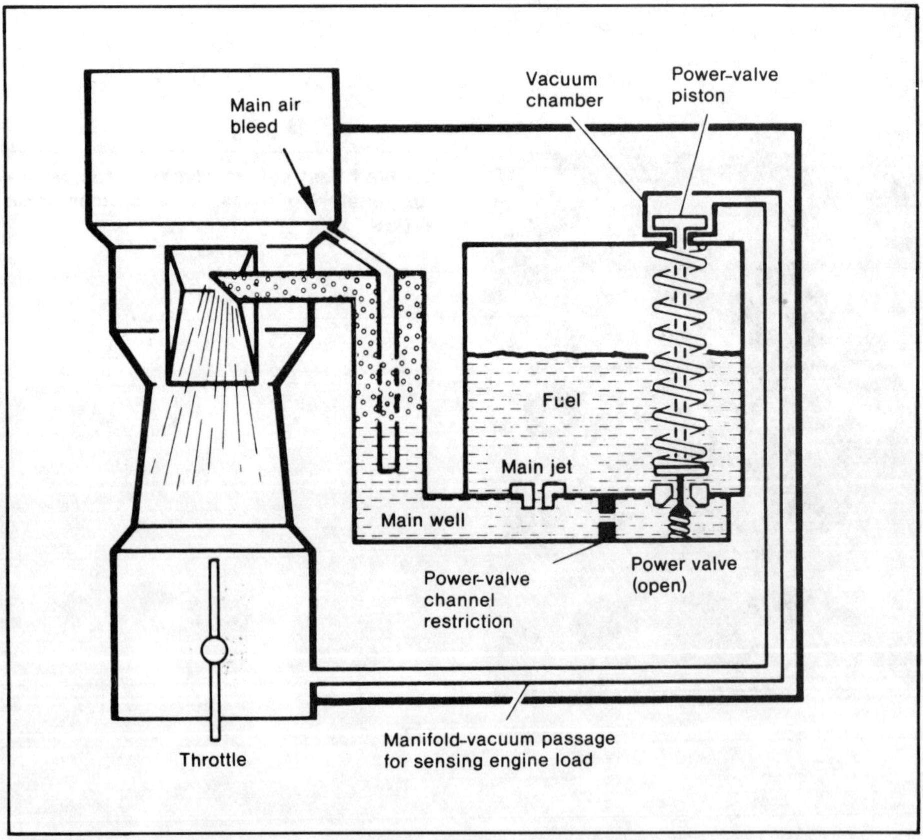

Typical power system. Additional fuel is supplied by the power system under high-load conditions as the power-valve spring overcomes low manifold vacuum.

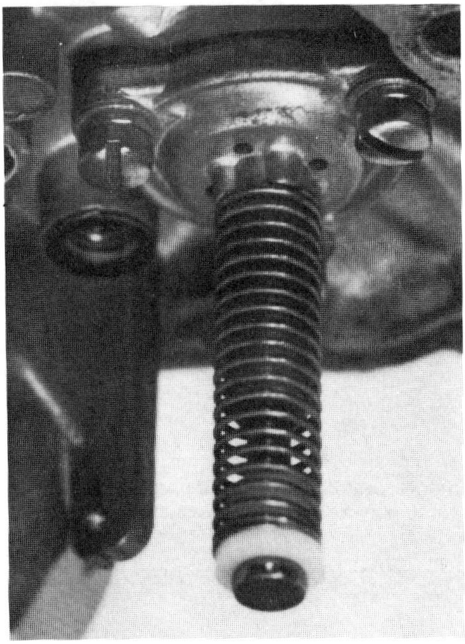

Power-valve diaphragm assembly as installed in the air horn. Vacuum closing force is applied to a diaphragm fastened to the stem. Spring on the stem provides the opening force. The white plastic spacers on the stem allow slight changes in power-valve timing. Removing spacers delays the opening to a lower manifold vacuum—larger throttle opening.

During part-throttle or cruise conditions, the object is to conserve fuel, so excess air is supplied to the engine to consume as much of the fuel as possible. The result is increased fuel economy and reduced exhaust emissions. Typical fuel/air ratio for economy—except for feedback systems which are covered later—is about 0.06 to 1 (16.7 to 1 air/fuel ratio).

At high-load conditions, the object is maximum power, so we want to consume all the air. This is done by supplying a little excess fuel. Typical fuel/air ratio for power is about 0.07—0.08 to 1 (14.3—12.5 to 1 air/fuel ratio).

During high-speed or heavy-load operation when manifold vacuum is low—near atomospheric—the carburetor automatically provides additional fuel through the power system.

Power valve at the bottom of the fuel bowl.

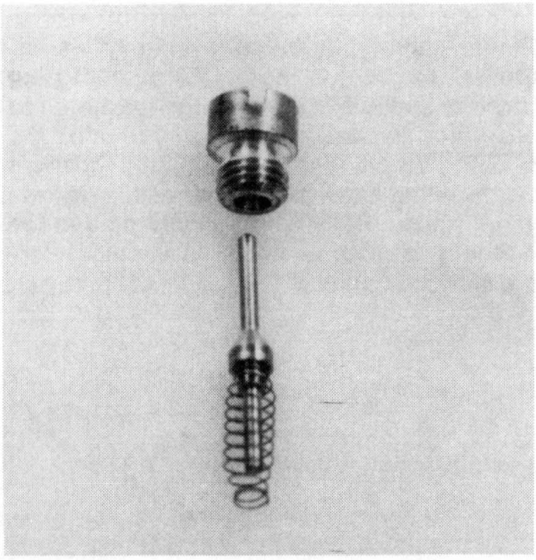

Power valve removed. Spring holds valve closed under normal circumstances.

Single-piece power valve operates the same way as three-piece unit.

A passage through the main body and air horn *communicates* manifold vacuum from a port below the throttle plates to a cavity above the power-valve diaphragm. This diaphragm is part of an actuating assembly that includes a stem, spring and housing. The assembly is secured to the bottom of the air horn with three screws.

The power valve itself threads into a cavity at the bottom of the fuel bowl. It consists of a seat, stemmed valve and a return or closing spring. Sometimes these three items are combined in a single assembly.

Manifold vacuum applied to the top of the power-valve diaphragm keeps it in the up position. When manifold vacuum drops to a given value—usually about 6''Hg—the spring overcomes the vacuum force and the stem moves down to open the power valve.

The opening point can be changed with a different spring or by changing spring preload or installed height by adding or subtracting spacers on the stem. Adding spacers causes the valve to open at a higher vacuum. Removing spacers causes it to open at a lower vacuum.

The power valve is simply a switch that opens at a predetermined manifold vacuum. It does not meter the fuel—this function is handled by a pressed-in restric-

Power System (Primary)

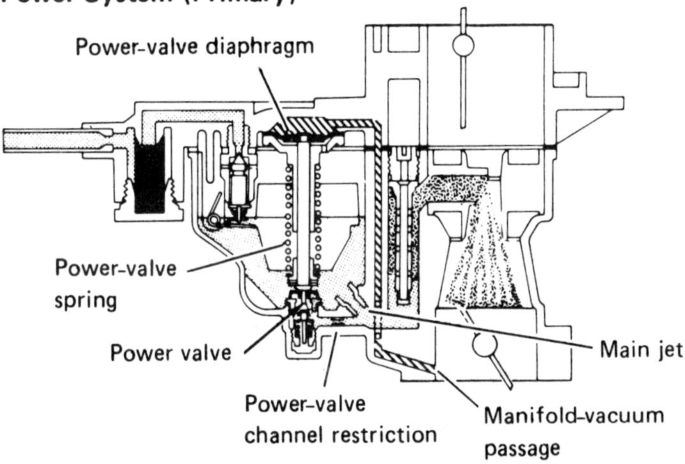

tion. When the diaphragm stem unseats the power valve, fuel enters a cavity below the valve, passes through a channel containing the power-valve-channel restriction (PVCR) and enters the main well.

The power valve affects the main metering system of the *primary side* only. Most of these carburetors have another, nonmechanical power system on the secondary side. I'll explain this in the section on the secondary system.

5 ACCELERATOR-PUMP SYSTEM

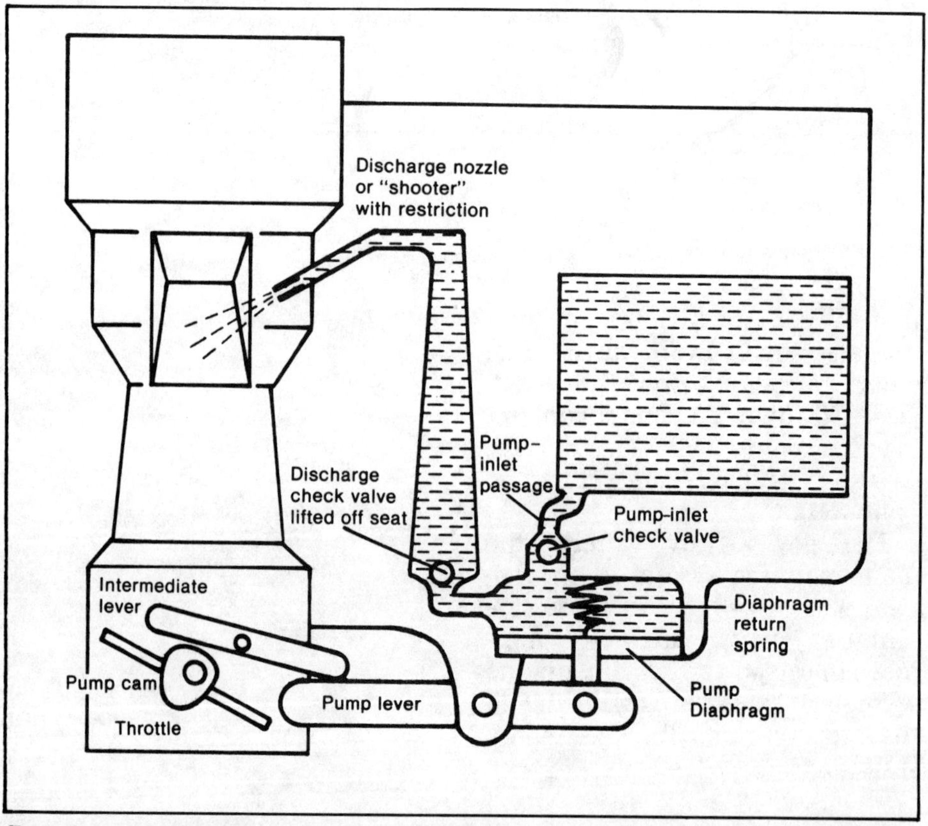

Typical diaphragm-type accelerator pump. Fuel is drawn into the pump from the float bowl past the check ball. When the accelerator linkage actuates the pump arm, the inlet check valve closes. Fuel is forced through the discharge passage then past the discharge check valve and out the nozzle. Although 5200 accelerator-pump system is arranged differently from that shown, operation is identical.

Cam (arrow) actuates accelerator-pump lever as throttle is opened. Lever is cut away for illustration.

The accelerator pump has two functions:

1. To make up for the lag in fuel delivery when the throttle is opened and more air rushes in.

2. To make up for fuel condensing onto the manifold surfaces when the throttle is opened suddenly at low engine speeds.

Because air is lighter—has less mass—than fuel, airflow responds more quickly than fuel flow to changes in throttle setting. Whenever the throttle is opened, the engine instantly gets more air through the larger throttle opening. But fuel flow in the main system doesn't respond as quickly to match this increased airflow. The accelerator pump mechanically injects the additional fuel needed until the main system restores the correct fuel/air ratio.

This lag in fuel flow is compounded if the throttle is opened suddenly at low rpm. Under these conditions there is a large instantaneous drop in manifold vacuum.

A high manifold vacuum keeps the mixture well vaporized. As manifold vacuum drops, fuel condenses out of the mixture and collects on the intake-manifold surfaces. Thus, the mixture instantly leans out just when the engine needs more fuel. Unless more fuel is immediately added the engine hesitates or stumbles.

This additional fuel is especially important with big-port heads and

Removable accelerator-pump nozzle. Only the primary side, left, is drilled.

Accelerator pump in middle position 2. Moving pivot pin to hole 1 decreases pump capacity, position 3 increases it. Some models have only two holes.

Pump-lever housing should be removed when changing positions. Support pump casting while removing and inserting pivot pin to prevent breaking the casting.

Remove pivot pin by driving it out with a small punch from the non-knurled side. Replace by driving from the knurled side.

manifolds, and manifolds with large plenums (volume below the carburetor) because there is less fuel/air velocity and more surface area where fuel can condense.

To counteract this low-flow, low-vacuum period, the accelerator pump injects—"shoots"—fuel into the throttle bore the instant the throttle is opened. This "shot" must be carefully metered because the low mixture velocity will cause much of the fuel to drop out before it reaches the cylinders. Duration of the shot must also be tailored to provide a "cover-up" long enough to allow main-system flow to be established.

This function—mechanical injection—is most important when the carburetor is higher in capacity than ideal for the application, or when a mechanical secondary throttle is used, as on 5200s.

Accelerator-Pump Actuation— When the throttle linkage actuates the pump lever, pressure forces the pump-inlet check valve onto its seat. This prevents fuel from flowing back into the fuel bowl. Fuel flows from the pump through a diagonal passage in the main body. Fuel pressure then raises a discharge check ball off its seat and fuel sprays into the venturi from the discharge nozzle.

As the throttle closes, the pump linkage moves toward its original position. A spring forces the pump diaphragm back. The pump inlet

Pump-discharge components removed—steel balls are discharge check valves. Second ball is for weight and not always used.

valve opens and the pump refills. When pressure drops in the pump passages, the pump-discharge check ball closes to prevent air from entering the pump chamber as it refills. This check valve also prevents airflow past the discharge nozzle from pulling fuel out of the pump passage.

Pump capacity is measured in cubic centimeters per ten full strokes of the throttle lever. It is a function of diaphragm displacement, governed by the pump-cam profile. Capacity can also be altered by changing the pivot point of the pump lever. See the adjustment section.

The duration of the pump shot is controlled by the size of the discharge nozzle. A small hole spreads the shot out over a greater period of time, a larger one has the opposite effect.

The pump discharges into the primary side of the carburetor only, but the pump shooter can be modified to divert part of the shot into the secondary side if desired. The drilling is a little tricky.

In most 5200-series carburetors fuel vapor generated in the pump chamber during *hot soaks*—hot engine shutdown—is vented into the fuel bowl through a restriction.

Accelerator-pump discharge nozzles are targeted so the pump shot "breaks" on the booster venturi. Photo taken with a four-barrel carburetor on air box shows how shot is pulled toward lower edge of booster venturi by air streaming through the carburetor. Bubbly fuel flow from main discharge nozzle issuing from tail of booster indicates air is already mixing with the fuel.

Accelerator-Pump System

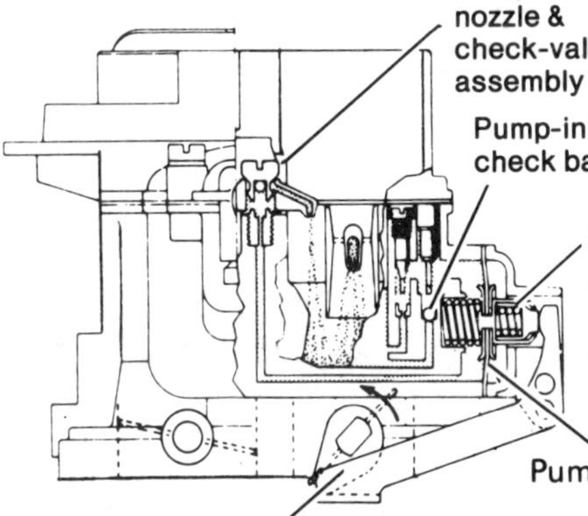

The Holley 2300 Carburetor Handbook is another HPBook covering specific Holley carburetors. The model 2300 two-barrel has been used as original equipment and aftermarket replacement since 1957.

6 SECONDARY SYSTEM

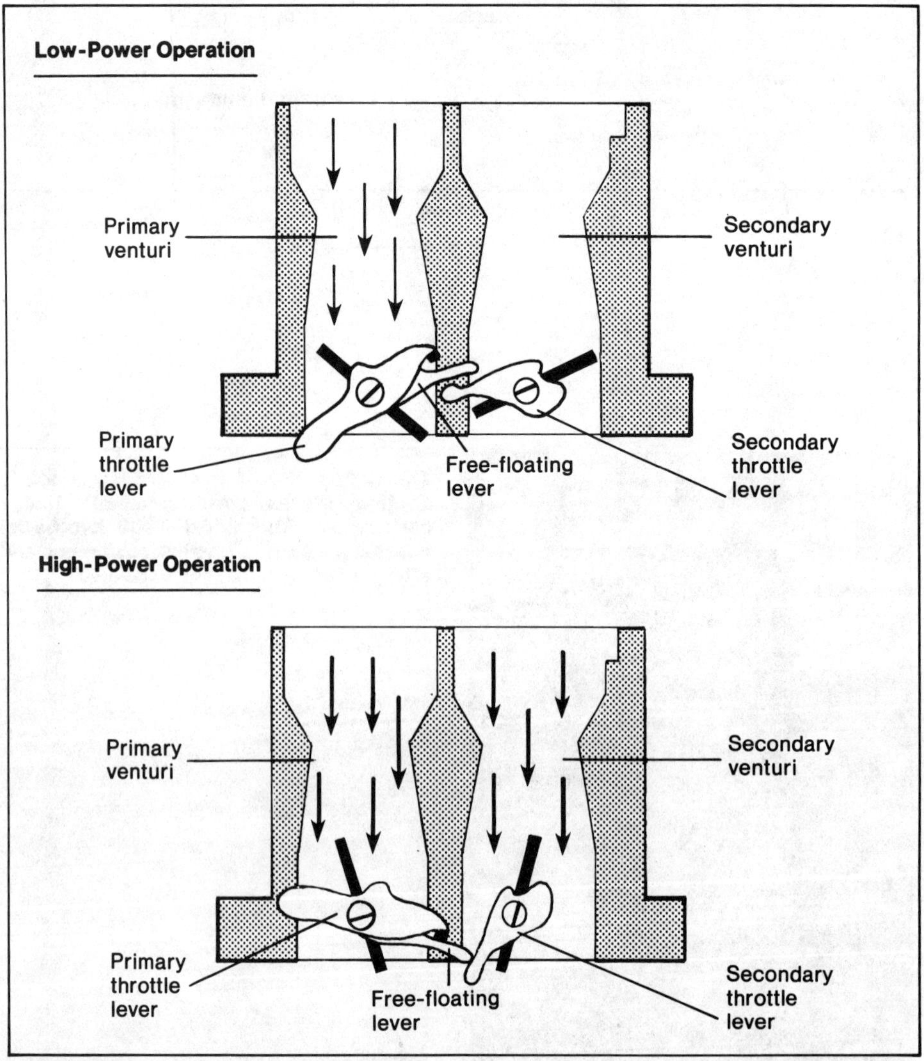

Primary system provides high venturi velocity to increase main-metering signal. Secondary system allows increased power at high rpm. Mechanical operation is similar to some Holley high-performance four-barrels.

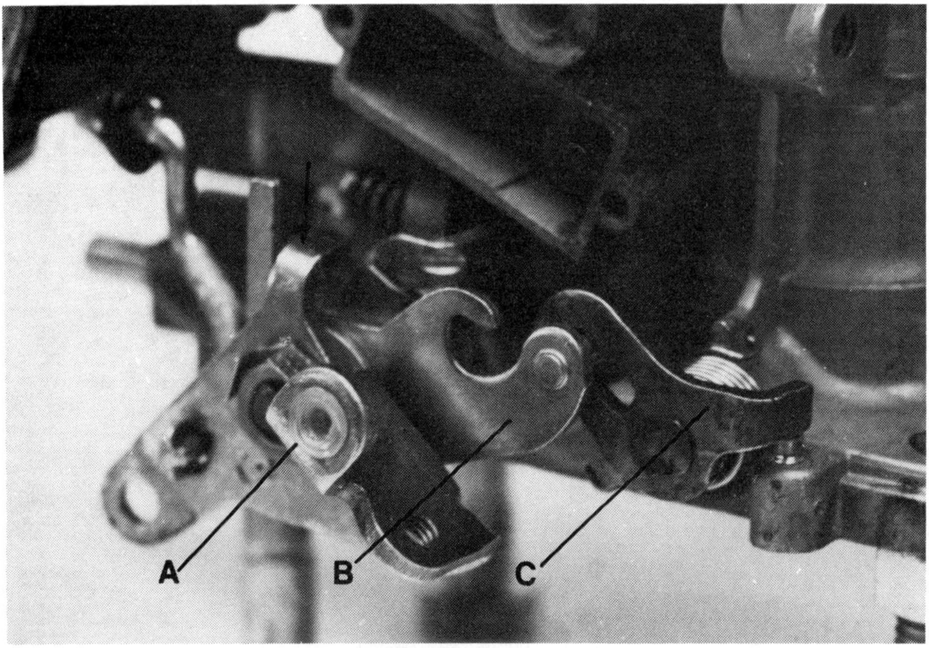

Lever A rotates clockwise to open the primary throttle. At a given primary angle—usually about 40°—tang (arrow) contacts free-floating lever B and operates lever C in a counter-clockwise direction to open the secondary throttle.

Earlier I stated the 5200 is a *staged* two-barrel. By this I mean it is essentially two carburetors mounted on the same flange. One half of this carburetor—the primary side—operates all the time.

All 5200-series carburetors use a mechanically operated secondary. When the primary throttle reaches about 40° of travel, the throttle lever contacts a free-floating lever on the primary-throttle shaft. This lever opens the throttle of the other half of the carburetor, the secondary.

The levers are designed to open the secondary throttle at a faster rate, so both primary and secondary throttle plates reach their wide-open positions at the same time.

Why a secondary system? Well, it's a way of having your cake and eating it too—all it costs is money.

A staged carburetor has a reasonably small primary venturi to take advantage of the good things brought about by high venturi velocities—strong metering signals and good mixing, atomization and vaporization. At the same time, having a second carburetor in parallel—the secondary—provides the added flow capacity necessary to raise the power output of the engine.

On 5200s, sizing is such that approximately 45% of the total fuel/air mixture passes through the primary side at wide-open throttle.

Mechanical secondaries have the advantage of letting the driver

Secondary side has a velocity-sensitive enrichment system. Arrow points to fuel feed in the bowl.

control secondary opening. Also, a lower airflow restriction offers a torque gain in the middle speed range of wide-open throttle.

The secondary side has a main system very similiar to that on the primary side. The earlier description of main-system operation applies. Some of these carburetors have a secondary idle system while others do not. None have an adjustable secondary idle mixture.

Secondary Power System—Most 5200-series carburetors use a velocity-sensitive power system on the secondary side. This non-mechanical system is operated by airflow in the secondary venturi.

At low rpm, 4-cylinder engines have the characteristic of pulsating heavily at wide-open throttle. What actually happens is the air moves back and forth through the carburetor. As the air moves in both directions through the boost venturi, it picks up fuel from the main system. This creates an enriching effect that diminishes as engine speed increases and the pulsations smooth out.

This is the same phenomenon that produces *fuel standoff*—a small cloud of fuel hovering over the carburetor—often seen on racing engines.

When airflow past a discharge port in the secondary venturi reaches a specific point, fuel is drawn from the bowl through a vertical passage containing a metering restriction. At the top of the passage, air from a separate air bleed is mixed with the fuel. This mixture flows through a cross passage containing a discharge port leading to the air horn.

The restriction in the vertical fuel passage determines the amount of enrichment, while the size of the air bleed affects both

Restriction (arrow) in vertical channel helps to shape fuel-delivery curve.

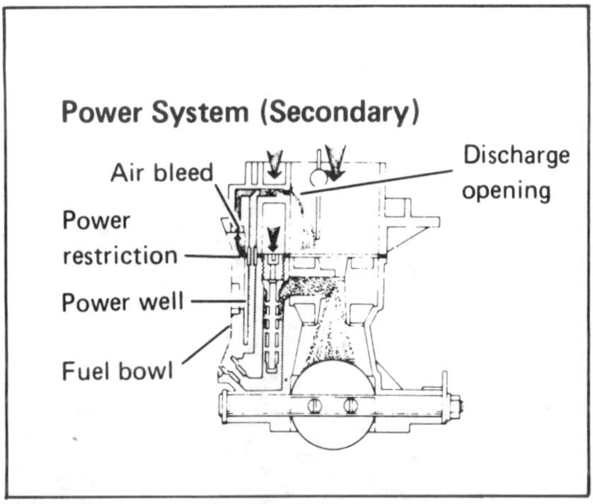

Air bleed (arrow) for velocity system.

Arrow indicates mixture-discharge port into air section. Locating it high in the air horn ensures operation only at high rpm as system requires significant fuel "lift."

the amount of enrichment and the airflow at which the system begins to flow.

Without this additional power system, wide-open throttle at low rpm would create a rich mixture that would lean out as engine speed increased. But an engine needs a fairly constant fuel/air ratio, regardless of speed. This special system allows the main metering system to be tuned to eliminate low-speed "richness" caused by the pulsing, but adds fuel as airflow increases and the pulsations diminish.

HPBooks publishes a number of books on Holley carburetors. The Holley 4150/60 Carburetor Handbook covers the most popular series of four-barrel carburetors ever produced by Holley.

7 CHOKE SYSTEM

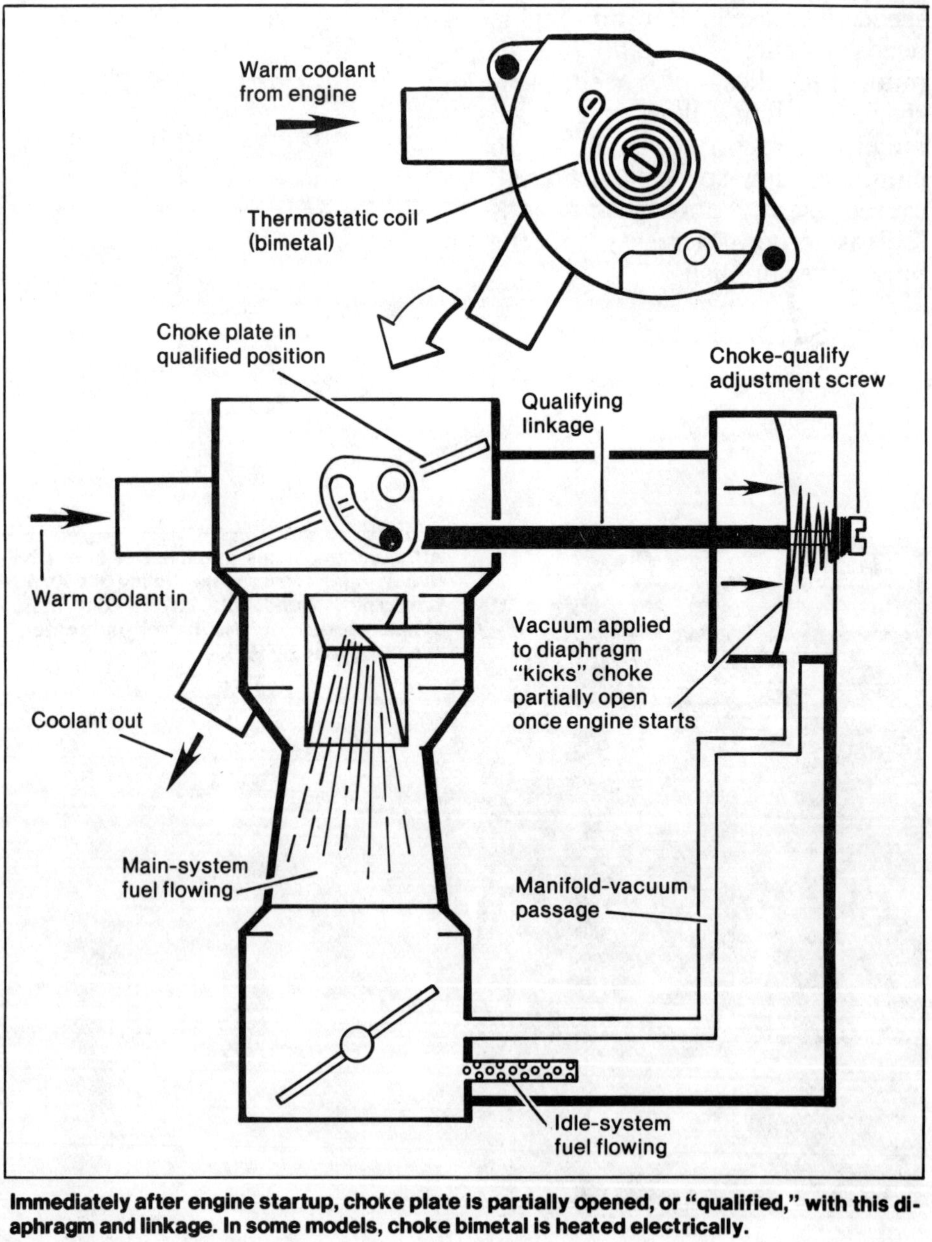

Immediately after engine startup, choke plate is partially opened, or "qualified," with this diaphragm and linkage. In some models, choke bimetal is heated electrically.

Coolant-heated integral choke.

The choke must perform several functions. First, it must provide an extremely rich mixture to start a cold engine. Closing the choke plate in the air horn creates a strong signal at the discharge nozzle and provides an extremely rich mixture.

A much leaner mixture is required once the engine has started, but still quite a bit richer than normal. As the engine warms the choke gradually opens—leaning the mixture to a normal level.

Due to higher internal friction, extra power is needed to idle a cold engine, so a *fast-idle cam* is mounted to the choke-control shaft. The *fast-idle lever* stops against this cam during choking. A rod attached to the fast-idle lever holds the throttle open slightly to give a greater throttle angle and a faster idle to aid atomization.

The choke plate is offset on its shaft and spring-loaded toward the closed position, so increased airflow opens the plate to provide a leaning effect.

Once the engine has started, the choke plate is moved to a partially open or *qualifying* position by a device operated by manifold vacuum. The qualifying is done by a diaphragm in the choke housing.

All these carburetors have integral automatic chokes. The integral choke has a flat, spiral-wound bimetal spring in a housing attached to the main body. All the functions described here are incorporated into the choke housing mounted to the main body. The resultant movement is transferred to the choke lever through a rod from the lever on the choke-housing shaft.

Two basic methods are used to heat the choke bimetal on 5200s.

Some have a portion of the engine coolant routed through a housing adjacent to the bimetal housing. This allows the bimetal to "see" coolant temperature as

Choke-qualifing housing. See Repair and Adjustment section for calibration.

Bimetal located next to its engine-coolant heat source.

All-electric integral choke.

All-electric automatic choke introduced in 1977 on Pontiac and AMC vehicles. Attachment is basically the same as for the hot-water choke.

an indicator of engine temperature. The bimetal reacts to this heat and gradually opens or closes the choke accordingly.

An alternate heating method is electrical. In this case voltage is applied to a heating unit near the bimetal. When the engine is started, the circuit is activated and the unit heats up. As the temperature increases, the bimetal gradually opens the choke.

The heating unit is a small disc called a *Positive Temperature Coefficient* (PTC) unit. The device raises its electrical resistance automatically as temperature increases, thereby lowering current flow.

Both of these methods are effective but the electrical choke offers a simpler installation. Certain versions use a combination of both.

On most 5200s, a tab on the fast-idle lever contacts the fast-idle cam near wide-open throttle to open the choke. Called *unloading,* this clears the engine in case it gets loaded, or flooded, with fuel due to excessive cranking.

Most of these carburetors have a choke plate on both the primary and secondary sides. This allows some enrichment to the secondary side if the secondary throttle is opened during warmup.

Opening the secondary throttle during warmup should be avoided. This causes excessive enrichment. The additional fuel washes the oil off the cylinder walls causing excess piston-ring and bore wear. Some aftermarket-replacement versions do not have a secondary choke.

Choke System

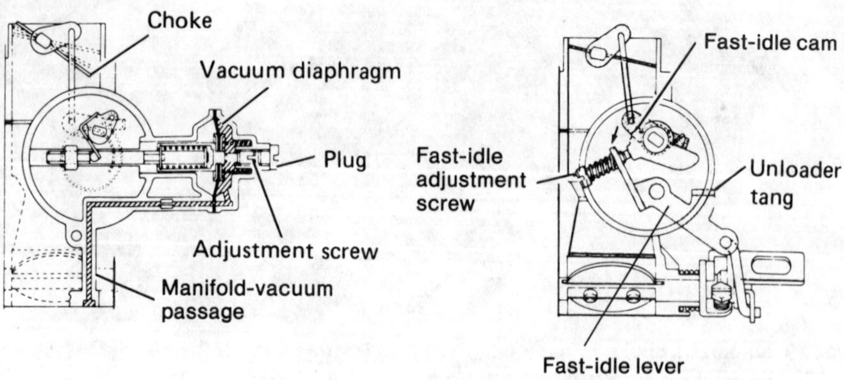

Most 5200-series carburetors have choke plates on both the primary and secondary sides.

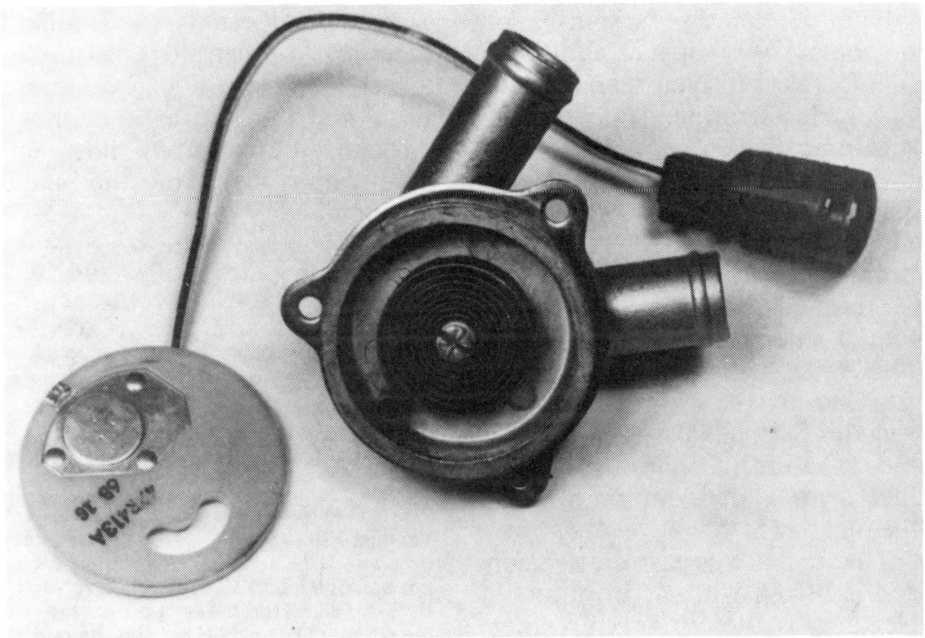

Combination of coolant- and electric-heated choke, known as electric-"assist" type.

8 DECELERATION AND FEEDBACK SYSTEMS.

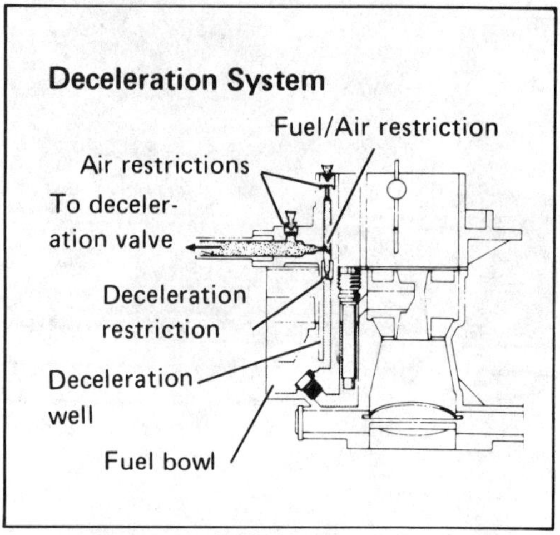

DECELERATION SYSTEM

The deceleration system used on some 5200s supplies additional fuel to an excessively lean mixture during deceleration. This prevents misfiring, which would increase hydrocarbon (HC) emissions. This system operates only when manifold vacuum is higher than normally experienced at idle.

During deceleration, high manifold vacuum passes through a *deceleration valve* and vacuum line attached to the carburetor. This vacuum is applied to a fitting on the carburetor air horn, and draws fuel from a deceleration well in the fuel bowl.

The fuel moves past a restriction and is mixed with air from an air bleed. This fuel/air mixture then moves through another metering restriction and into the cross passage connected to manifold vacuum. Another air restriction in this cross passage adds even more air to the mixture as it is drawn through the vacuum line and deceleration valve into the intake manifold.

Vacuum fitting for deceleration valve used on some 5200s. When used on an engine not equipped with a deceleration valve, the 0.25-in.-diameter brass tube on the air-horn/cover assembly should be plugged.

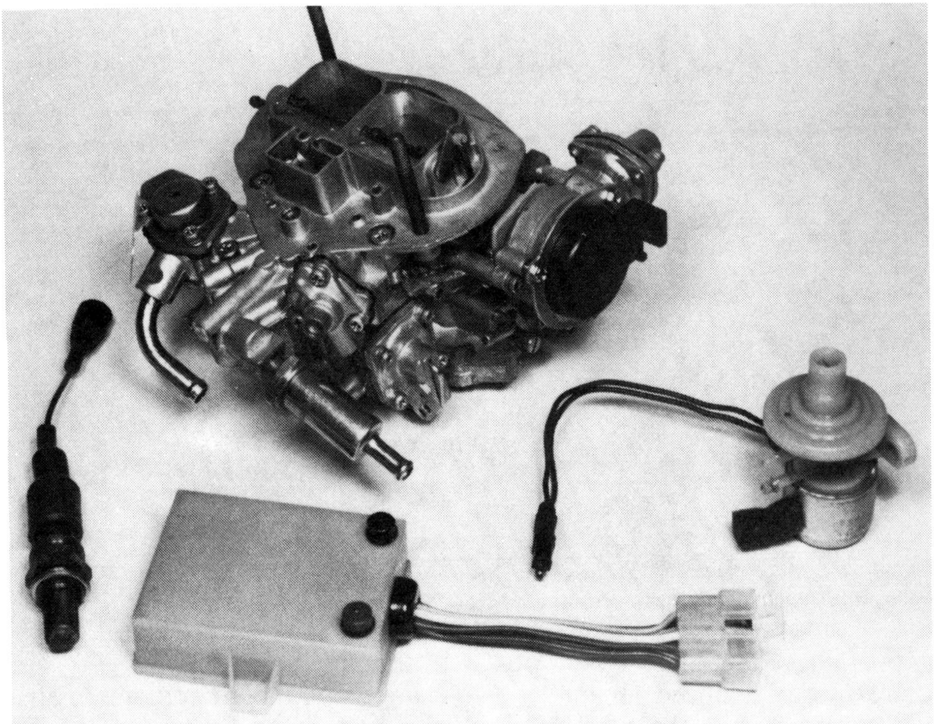

Major components of vacuum feedback system. Starting at the left and going clockwise are the oxygen sensor, carburetor, solenoid and vacuum control valve, and electronic control unit.

FEEDBACK CARBURETION

In general, gasoline engines produce three emissions—hydrocarbons (HC), carbon monoxide (CO) and oxides of nitrogen (NO_x). The amount of each emitted by a vehicle under controlled test conditions is limited by law.

Emission standards have become more and more difficult to achieve over the years, and things have become further complicated by the Corporate Average Fuel Economy (CAFE) standards. Turn to page 54 for an explanation of CAFE standards. Changes made to satisfy one set of standards may negatively affect the other.

In 1975 we saw the introduction of the exhaust catalyst as a method of handling HC and CO. The original catalytic convertors are *oxidizing* catalysts—they encourage both compounds to take on more oxygen. Carbon monoxide is changed to carbon dioxide (CO_2), which is harmless, and hydrocarbons are converted to water (H_2O) and carbon dioxide.

Oxides of nitrogen (NO_x) were lowered by engine modifications that lowered combustion temperatures. Most of the improvement was done by lower compression and exhaust-gas recirculation (EGR), which dilutes the intake charge with exhaust gas.

Holley 6500 vacuum feedback carburetor. Arrow indicates control diaphragm.

Further reductions of NO_x require more advanced technology, which brings us to the *feedback* or *closed-loop* system. If the concentrations of HC, CO and NO_x can be held in proper balance they can all be treated by a single catalyst—hence the term *three-way catalyst.*

What happens is really quite simple to describe—oxygen is released from the NO_x and combined with the HC and CO. The NO_x becomes free nitrogen (N_2)—an inert gas under normal conditions—and the HC and CO become water and carbon dioxide.

The trick is maintaining the correct balance. The ideal combination occurs near the chemically correct or *stoichiometric* air/fuel ratio—approximately 14.7:1.

Maintaining the emissions balance is the major reason for the closed-loop or feedback system. It determines the fuel/air ratio from a sample of the exhaust gas, and changes the fuel metering to maintain the balance.

Here's how it works. In addition to the three pollutants mentioned, the concentration of oxygen also varies with fuel/air ratio in a predictable manner. Consequently, the amount of oxygen in the exhaust can be used as a guide to adjust the fuel/air mixture.

An oxygen sensor is in the exhaust upstream of the catalyst. The sensor puts out a low-voltage signal.

Vacuum Feedback System—On the vacuum feedback system, (Holley 6500) this signal is fed through an amplifier to a switch. This switch controls a valve that channels manifold vacuum to a special diaphragm on the carburetor. The diaphragm mechanically controls a bleed to a revised power valve.

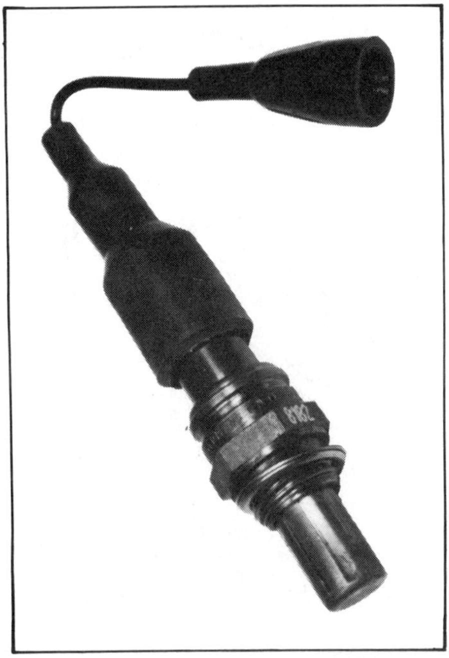

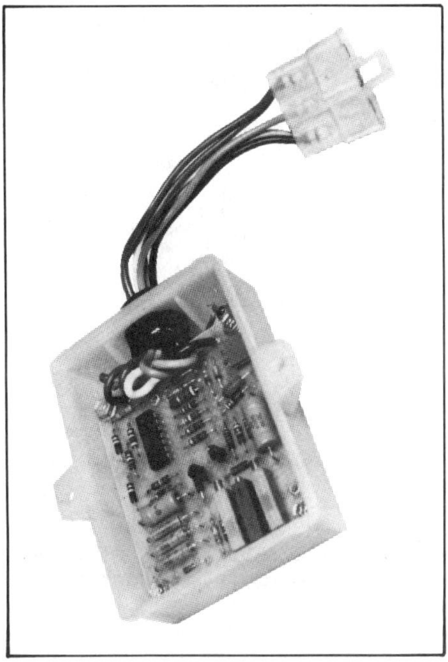

Bosch Lambda sensor is in the exhaust system, upstream of the catalytic convertor. Output goes to electronic control unit (right). Much development work has been done to produce compact and reliable oxygen sensors.

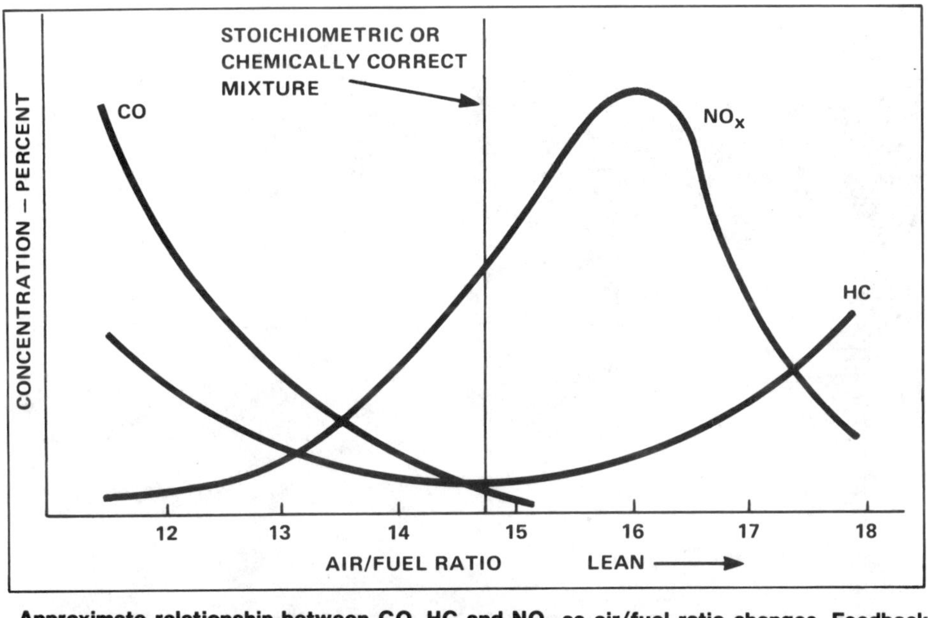

Approximate relationship between CO, HC and NO_x as air/fuel ratio changes. Feedback system is designed to keep the air/fuel ratio close to the stoichiometric line.

Power valve for vacuum-feedback system is in standard position. Unlike a standard power valve, vacuum-feedback power valve actually meters fuel rather than simply opening or closing.

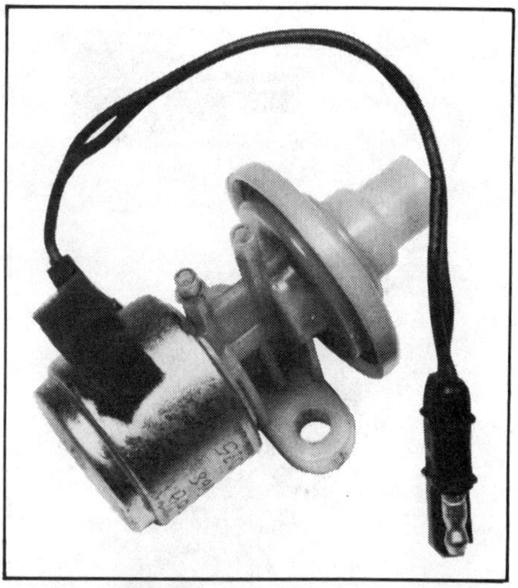

Vacuum control unit. Hose connections channel manifold vacuum to control diaphragm on carburetor.

Diaphragm and control valve disassembled. Valve controls bleed to power-valve diaphragm to vary fuel/air ratio.

This carburetor is for the 1981 Chevette. The duty-cycle solenoid can be identified by the electrical lead at the left. The plastic connector at the lower right is to a throttle-position sensor, which supplies input to the electronic control unit. The rest of the carburetor is pretty much conventional. The duty-cycle solenoid can be taken out by removing the screws at each side of the tower. (arrows)

Vacuum Feedback Carburetor

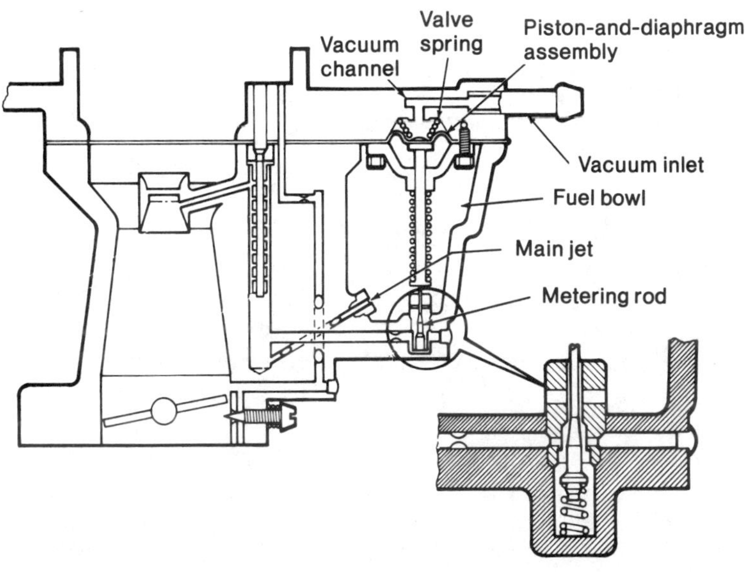

Courtesy Ford Motor Company

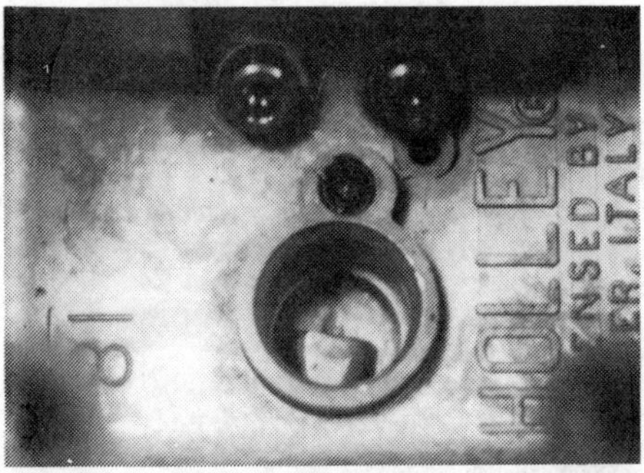

Modified power-valve cavity accepts the bottom of the solenoid valve. A channel leads into the primary main well. Conventional power system is unnecessary with the feedback system.

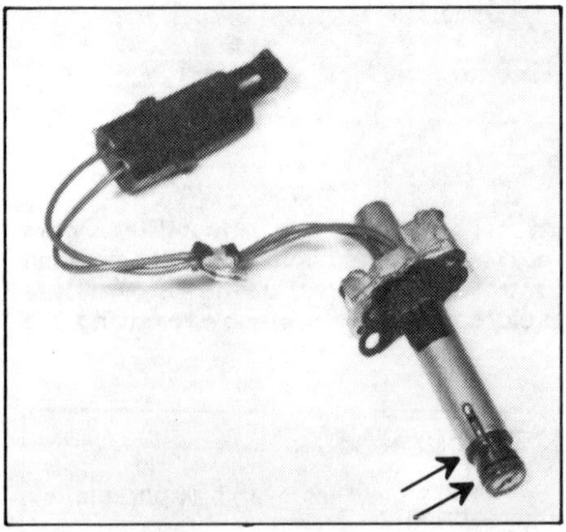

Duty-cycle solenoid removed from the carburetor. Actual fuel-control valve is at the bottom. The two O-rings (arrows) prevent fuel from flowing around the valve.

The vacuum feedback system (Holley 6500) was introduced on the 1978 Ford 2.3-liter engines for California. Application was extended nationwide on the same engine in 1981.

Electronic Feedback System—A new version of the feedback carburetor was introduced in the 1980 model year for Chevette (Holley 6510) and Omni/Horizon (Holley 6520) for California only. Application was extended nationwide for 1981.

With this system, the signal originates at the oxygen sensor as before, but the electronic unit is different. It generates a 10 Hertz (10 cycles per second) signal. This signal is sent to a solenoid-controlled valve in the carburetor.

This *duty-cycle solenoid* opens and closes at the same frequency. The percentage of time the valve remains closed—the duty cycle—during each cycle varies, depending on the signal from the control unit.

For example, under certain conditions the valve might be open 0.06 second, closed 0.04, open 0.06, closed 0.04 and so on. If the oxygen sensor indicates this is too rich, the timing might reverse to open only 0.04 and close 0.06. Either way, each cycle, (open *and* closed) is 0.10 second.

On some carburetors this valve operates on the main system only by opening and closing a restriction to supplement the main jet. Other carburetors have valves that also open and close an auxiliary air bleed affecting the idle mixture as well.

Mixture is controlled by the length of time the valve is held open or closed. The signal is constantly being altered as the oxygen content in the exhaust is monitored.

When manifold vacuum drops to a given value under heavy load, a sensor signals the control unit. The valve then automatically goes to its richest metering point, much like a conventional power valve.

Electronic Feedback Carburetor

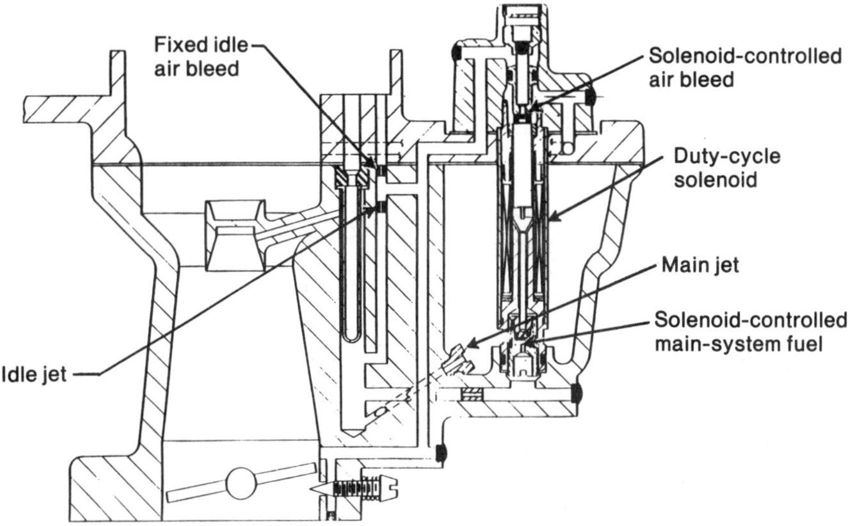

Open-Loop System—The technology used in the closed-loop system can be used to improve fuel economy. With this in mind, Holley introduced a driver-controlled *open-loop* system.

In this system, the feedback system is replaced by a computer and driver-controlled switch. The solenoid-controlled air bleed is eliminated, so the duty-cycle solenoid affects only fuel flow.

The solenoid is controlled from an 11-position switch that changes the duty cycle—the percentage of time the solenoid is closed—from maximum rich to maximum lean. On the 5200 series the duty cycle varies from 30%—maximum rich—to 80%—maximum lean. This is equal to about 12 jet sizes.

Complete 5200-series open-loop kit includes fuel-metering control (top), computer and all necessary hardware and wiring. Computer monitors a number of functions, including average and instantaneous fuel consumption. Kits are available with or without the computer.

9 FUEL ECONOMY

The major fuel crises we experienced a few years ago and the continually increasing cost of gasoline have made us all more aware of fuel economy. U.S. Federal law now dictates the fuel economy of new vehicles, known as Corporate Average Fuel Economy, CAFE.

Based on manufacturers' fleet averages, a minimum of 18 miles per gallon (mpg) was required for the 1978 model year, increasing to 27.5 mpg by 1985. These numbers are not just targets for the automotive companies—they are law.

The basis of measurement is the Environmental Protection Agency (EPA) emission tests, also used to calculate "city" and "highway" economy. The number referred to in the law is a composite of the two, weighting city economy at 60% and highway at 40% on a *sales-weighted* basis. Sales weighted simply means that the *number* of vehicles sold has an effect on the average. The largest-selling model of the manufacturer has the largest influence.

While the computation is a bit complicated, the impact is obvious. Take a look at the new cars being introduced and you can see the trend indicating how these objectives will be met. The result is total vehicle redesign. The major effort is toward reducing rolling resistance and weight, using smaller engines and improving vehicle aerodynamics.

Notice I didn't list carburetors or fuel systems as major areas of potential improvement.

Before going further, I'd like to talk about a bit of automotive mythology—"The 100 Mile-Per-Gallon Carburetor." The basic story goes: "The carburetor companies have carburetors capable of hundreds of miles per gallon, but they've sold out to the oil companies."

First off, let me tell you this: If there were such a thing, Holley would sell it to the public, not the oil companies, and make a lot more money in the process.

Secondly, it's the nature of the gasoline engine, not the carburetor, that prevents this kind of improvement. A carburetor can meter much leaner mixtures than

Fuel consumption depends on many aspects, not just the carburetor.

an engine can burn. What we really need is a 100-mpg engine.

There is a fixed relationship between fuel and air that represents a burnable mixture. That relationship is about 15—16 pounds of air for every pound of fuel. If the mixture is made leaner by adding more air, the burn rate in the combustion chamber slows down with a progressive loss of efficiency. This loss of efficiency causes a corresponding drop in fuel economy, as well as additional heat buildup.

Adding more fuel causes a similar drop in the burn rate and loss of efficiency, but the added fuel makes the fuel economy loss even worse. For maximum economy there is a given fuel/air relationship that must be maintained. This rules out the possibility of astronomical fuel-economy improvements with a given engine.

This is not to say nothing can be done in the way of carburetor design to improve economy. There are many ways to optimize fuel economy. These include:

1. Program fuel delivery to match engine requirements as closely as possible.

2. Mix the fuel with the air to optimize burning.

3. Distribute this fuel/air mixture equally to all cylinders.

4. Select carburetor size to obtain the best compromise be-

tween economy and performance.

This is what we do in carburetor design and development, but there are also things you can do to your carburetor to optimize economy. This does not include changing main jets, air bleeds, or making other calibration changes. Unless you're equipped and experienced for the job, the plain fact is *you'll probably lose ground.* You're much better off selecting a carburetor already calibrated for your vehicle.

MODEL 5200 ECONOMASTER CARBURETORS

Early in 1975 Holley decided to review several older carburetor models and incorporate some of these changes. At the same time some of these carburetors were redesigned to make them closer to true "bolt-on" replacements. The result is the Economaster line of carburetors, seven different models including the 5200 family.

The 5200 presented quite a challenge for a couple of reasons. First, it is currently being used in a great number of original-equipment applications and as a result is constantly being improved. Second, its staged design makes it a very sound carburetor in terms of fuel economy.

Staged operation is described earlier. In summary, the high primary velocity yields a strong main-metering signal. This provides good mixture preparation, better cylinder-to-cylinder distribution and allows somewhat leaner operation.

One change to the 5200 was the substitution of a double boost venturi for the standard single booster

Double boost venturi on primary side is one of the modifications found in the 5200-series Economaster carburetors. Double venturi cuts airflow slightly but improves distribution and increases metering signal.

on the primary side. This further amplifies the metering signal, allowing slightly leaner mixtures. There is a slight loss of airflow capacity—about 20 CFM—with the double boost venturi.

While the loss of power is indiscernible to the average driver, Economaster versions should not be used for competition.

The other change was a modification to the idle system that consists of a larger idle transfer slot in place of the smaller capacity holes or slots. This change increases flow through the idle system, further improving atomization to allow leaner operation.

The overall economy gain for the 5200 is between 5% and 10% when compared to the original carburetor. Testing followed EPA-type procedures.

MAINTENANCE

Besides the care and selection of the carburetor, there are many other things to take into consideration for maximum fuel economy. Probably the best thing you can do is to make sure your engine is in a good state of tune.

Beginning with your carburetor, here are some specifics:

1. Make sure the choke comes off, or opens, in a reasonable amount of time, and comes off *completely*. If it doesn't, check for binding or sticking linkage. Use carburetor cleaner to free things up if there's a problem.
2. Check the choke qualifying mechanism. If it sticks, clean it with carburetor cleaner.
3. Check all hose connections, especially those that supply vacuum to the distributor and other components.

Set ignition timing to the specification given on the decal or tag in the engine compartment. Also, check the dwell or point gap of a point-type distributor. Change the points and condenser regularly—

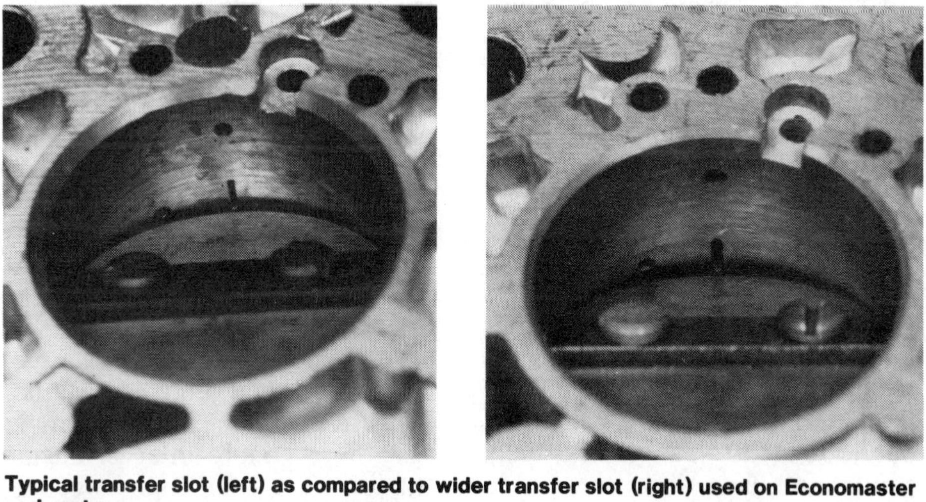

Typical transfer slot (left) as compared to wider transfer slot (right) used on Economaster carburetors.

about every 10,000 miles.

Make sure the vacuum- and centrifugal-advance mechanisms are operating properly. Check for bare wires and loose or broken connections, particularly with electronic ignitions. If any wires or hoses are hard or brittle replace them.

Change the oil and filter regularly to keep engine friction and wear down. Also, make sure the coolant level is maintained. Keep the drive belts at the proper tension. Overtightened belts waste energy and overload bearings.

Make sure all the emission components operate properly. In many cases there is a strong relationship between emissions and economy. One of the most important controls to check is the PCV valve. It fits in the large-diameter hose running from the base of the carburetor to the valve cover. Remove it and give it a shake. It should rattle.

CHASSIS

Check the suspension on a regular basis. Worn shocks and poor wheel alignment can hurt mileage. Brakes should also be considered. Dragging brakes not only waste energy, they wear out quickly.

Check tire pressures. Rolling resistance increases with low tire pressure, and this has a direct effect on mileage. In fact, it helps to run tire pressures a little higher than the auto manufacturers recommend. Don't go beyond the limits shown on the tire's sidewall, but around 30—35 psi is usually good. You'll sacrifice some ride quality, but you'll go farther on a gallon of gas.

DRIVING HABITS

There are a lot of positive things you can do just by the way you drive.

Letting the car warm up on cold winter mornings while you have a second cup of coffee will give you a nice cozy car, but it wastes fuel.

Avoid jack-rabbit starts. Hard acceleration lowers manifold vacuum and opens the power valve. This adds about 15% more fuel. Keep the power valve closed as much as possible. In fact, a simple vacuum gage is a tremendous driving aid. Most power valves open when manifold vacuum drops to about 6"Hg, so keep the gage reading higher when possible. It works!

Watch your speed. Power and fuel requirements increase rapidly with speed. It'll save you money in traffic tickets too.

Don't use the brake pedal as a foot rest. This'll really drain the tank and severely cut brake life.

Don't "exercise" the accelerator when you're cruising down the highway—on the gas, off the gas, on the gas. In addition to making passengers uncomfortable, it constantly fills and empties the accelerator pump—a real economy killer.

Finally, plan your trips. Make one longer trip take the place of several shorter ones. This is especially important in cold weather because the choke stays on longer and there's more internal engine friction in a cold engine, consequently economy is quite poor.

With all the emphasis on economy, there are all kinds of devices being offered with fantastic savings being claimed. In fact, if you installed the proper combination of these devices on your car, you'd have to make arrangements to drain your gas tank! The fact of the matter is, most of these offer little or no gain. In fact, some of them hurt fuel economy! We've tested a lot of these; so has the EPA with the same general results. All we can say is—*be cautious!* Look at the claims and make sure they make sense.

Keeping your air filter clean has a direct effect on miles per gallon. One of the biggest mileage "boosters" is normal maintenance.

JAPANESE VEHICLES

Most Japanese 4-cylinder engines are originally equipped with staged two-barrel carburetors of Japanese manufacture. As the number of these applications increased, Holley recognized the potential for the 5200 carburetor in the replacement market.

Model 5200s for Datsuns were introduced in mid-1980. Coverage goes back to the 1975 model year. Toyota, Chevrolet LUV trucks and Ford Courier trucks replacement carburetors—also going back to 1975—soon followed.

Actually, the basic carburetor design wasn't changed very much. Carburetor-to-manifold adapters are necessary because of different bore and stud spacing. Air-cleaner adapters are also required. All these adapters are packaged with the carburetor. The other major revision was to modify the throttle-linkage hookup to fit the existing vehicle linkage. All in all, the carburetors install quite easily, but the instructions must be followed closely because of the many connections.

A great deal of time was spent on each application to calibrate the carburetors to each engine. The results are a slight edge in performance and driveability over the original carburetor. Each carburetor meets the emission requirements for the recommended applications.

This particular carburetor is for the Datsun 2.0-liter engine. Choke (left) is all-electric. Throttle linkage and brackets are designed specifically for this direct "bolt-on" replacement.

Two-piece carburetor-to-manifold adapter. Bottom left photo shows assembled adapter.

Same carburetor as that shown at left with the air-cleaner adapter. This allows use of the original air cleaner. For off-road, use Holley high-performance air cleaner 120-106 which bolts on without the adaptor.

HOLLEY/VW CONVERSIONS

Although Holley makes 5200 carburetors for specific applications—particularly Japanese cars and trucks—a number of enthusiasts have adapted it to other engines. Several manufacturers currently market carburetor-conversion kits based on the 5200.

Much of the 5200's popularity is due to its adaptability to engines originally equipped with either a Weber or Solex staged two-barrel. Both of these carburetors use the same stud pattern as the 5200. Manufacturers that have used these carburetors include Alfa Romeo, Audi, BMW, Fiat and Volkswagen.

In addition, the variety of main and idle jets, main-well tubes, air bleeds and venturi sizes makes it possible to tune this carburetor for a large number of engines and displacements. All it takes is the patience to go *one step at a time.*

Fortunately for the VW owner the work has already been done. A number of manufacturers are currently marketing kits to adapt the 5200 to air- or water-cooled VW engines. Some also make dual-carburetor setups for high-performance or racing use.

Most of these kits include linkage, air filter, an adapter for the Zenith carburetor flange (if needed) and a properly jetted carburetor. Although this is by no means a complete list of kit manufacturers, I've included the names and addresses of three. You may want to check the VW enthusiasts' magazines for others.

Beck Enterprises
18527 S. Broadway
Gardena, CA 90248
(213) 538-9478
Conversions for air- and water-cooled, single- and dual-carburetor VWs.
Also conversions from fuel injection to carburetors.

Dee Engineering
Bugpack Performance Products
3560 Cadillac
Costa Mesa, CA 92626
(714) 979-4990
Conversions for air- and water-cooled, single- and dual-carburetor VWs.
Also conversions from fuel injection to carburetors.

ARC International
4788 Library Road (Route 88)
Bethel Park, PA 15102
(800) 245-4458
Zenith-to-5200 conversion for water-cooled VW.

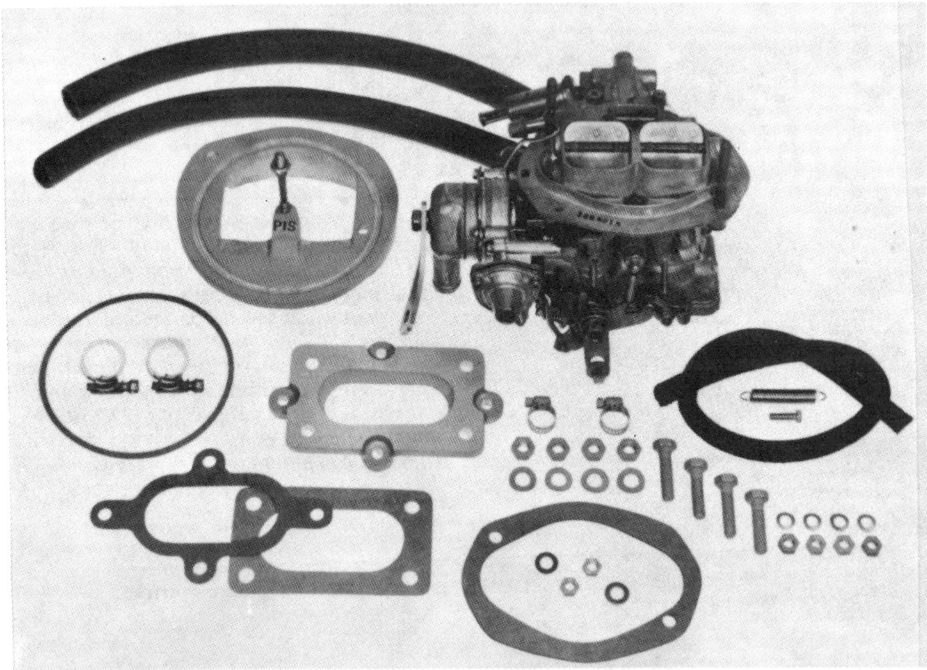

Beck Enterprises Progressive Induction System for Rabbit, Scirocco and Dasher (above), ARC kit (below) as installed. Kits include manifold and air-filter adapters, hose clamps and gaskets. Manifold adapter is needed only with Zenith carburetor, not Solex. Kits allow using stock air filter and stock-type hot-water choke.

10 CARBURETOR REMOVAL AND INSTALLATION

The first step: Read the instructions enclosed with your new Holley carburetor. The instructions cover adjustments and specifications for different applications.

Do yourself a big favor by working on a cold engine, and give yourself plenty of time. I usually look at the job, estimate how long it will take, double that time, then add an hour; it comes out pretty close! Remove the air filter and you'll get a good idea what kind of job is ahead.

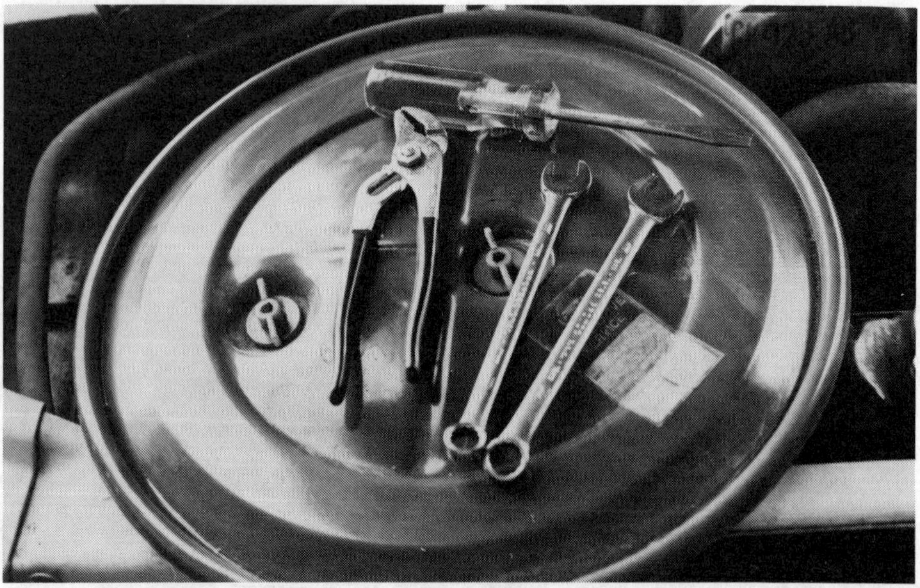

About all the tools you'll need to remove a carburetor, unless it uses a brass fuel line and flare nuts. Then you'll need a flare-nut, or line wrench. Air-filter top makes a handy parts and tool tray.

Masking-tape "flags" will identify vacuum lines and electrical connections. Some Polaroid snapshots or sketches also help.

There may be residual fuel pressure here when you remove the fuel line. Wrap the connection with a rag or shop towel to make sure the fuel doesn't spray when you pull the line off.

Manufacturers change the design of linkage retaining clips as often as you change your socks, so check them before you go pulling or prying on them. Also, if you drop a clip, find it before removing the carburetor. Otherwise it will fall unnoticed into the intake manifold (Murphy's law). Most carburetors have two or three rods and springs connected to the throttle lever, so you may want to flag these too.

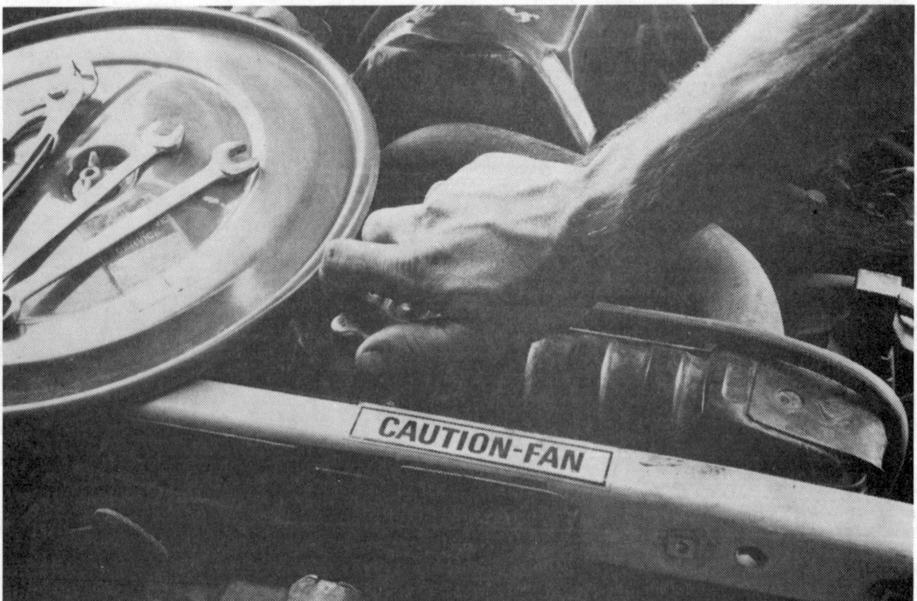

Before disconnecting a hot-water choke loosen the radiator cap to release any pressure. This is an absolute must on a hot engine.

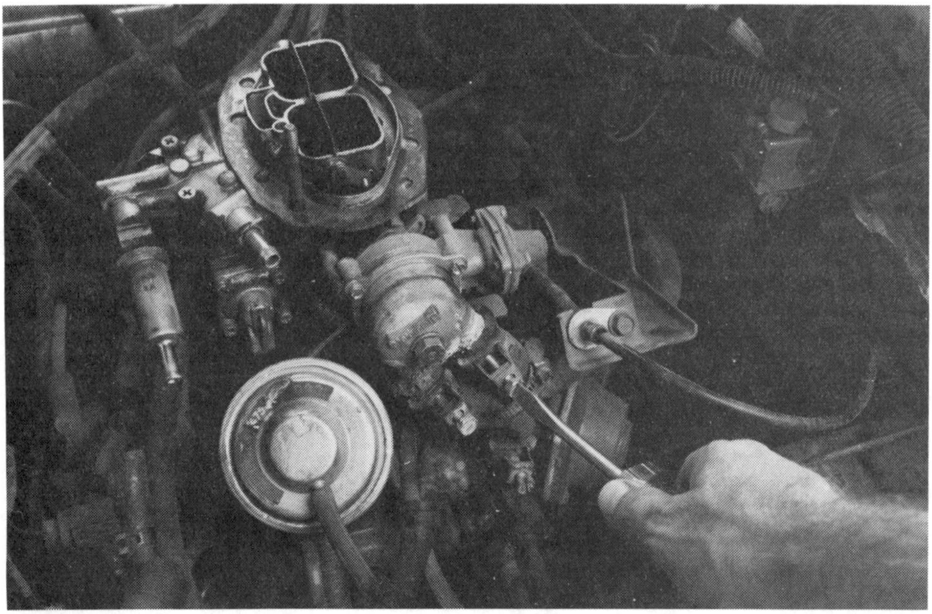

Be prepared to lose some coolant when you disconnect the hot-water choke. It's also a good idea to check the hoses and clamps here. Replace the hoses if they're old or brittle.

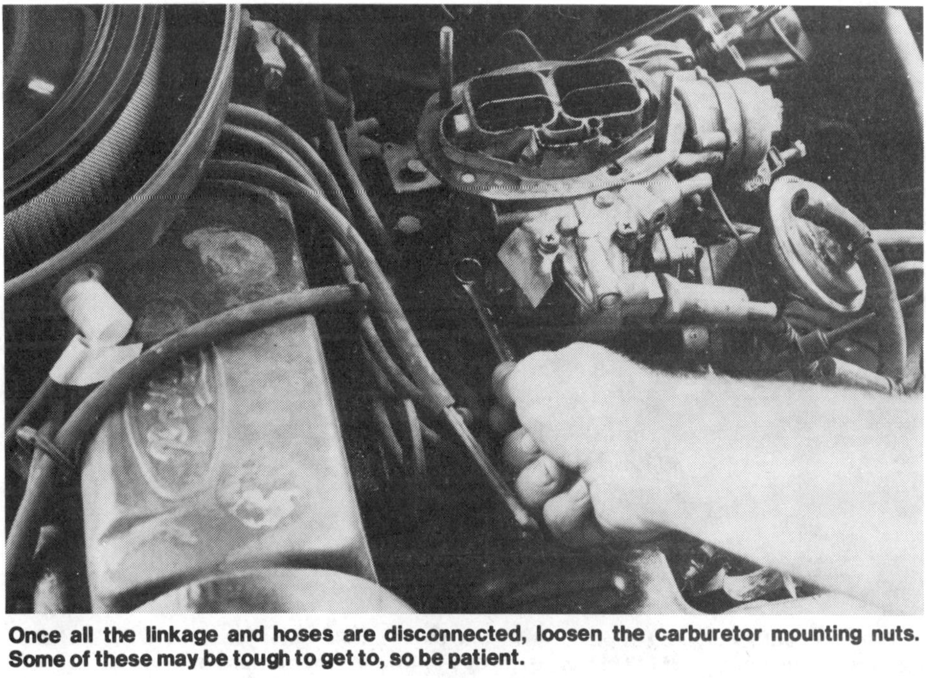

Once all the linkage and hoses are disconnected, loosen the carburetor mounting nuts. Some of these may be tough to get to, so be patient.

Before you lift the carburetor off, check to make sure everything is disconnected. I found this electric-choke-assist connector when the carburetor was reluctant to come off.

Keep an eye out for any dirt or parts falling into the manifold when you remove the carburetor. Any bits or pieces that fall in can be removed with a magnet or vacuum cleaner.

Holley supplies a new fuel filter, fuel line, clamps and a base gasket. Use them. A new air filter is also a good idea.

Place the carburetor on the intake manifold, using a new gasket. Don't bolt it down yet. Install any metal lines, like the fuel line, while you can still move the carburetor.

We're almost finished, so you can actually bolt the carburetor down. The most important things here are to tighten the bolts evenly, and don't overtighten them. Use a criss-cross pattern, and don't exceed 15 ft-lb. Although the fuel filter should be a dead giveaway, don't confuse the fuel line with the evaporative-canister line. The canister line connects here, near the air horn.

Start the engine. Remember, there's no fuel in the fuel bowl, so the engine's going to crank longer than normal until it fills. Set the idle according to specs for your car. Ballpark fast idle is 1700 rpm in neutral, curb idle about 700-850. With an automatic transmission, set curb idle in gear with the emergency brake on and the wheels blocked.

One final note. There's always the temptation to drive the car around with the air filter off, particularly when you've just replaced the carburetor and "just want to check it out." Don't do it! Engines don't like "fresh air" with all its dust and dirt. It takes a lot less than a small bird sucked into the carburetor to do severe engine damage.

CARBURETOR TROUBLESHOOTING

POSSIBLE CAUSE / PROBLEM	Stalling	Rough Idle	Flooding	Hot Start	Economy	Hesitation	Acceleration	Surge	Back Fire (cold)	Power	Stalling (cold)
Idle Adjustment	X	X		X	X	X		X	X		X
Idle Needles (Damaged)	X	X									X
Idle-Vent Adjustment	X	X		X							X
Fast-Idle Adjustment	X										X
Idle Passages (Dirty, Plugged)	X	X				X		X	X		X
Auto-Choke Adjustment									X		X
Choke-Qualify Adjustment									X		X
Choke-Rod Adjustment									X		X
Choke-Unloader Adjustment									X		
Metering Jets (Loose, Plugged)					X	X	X	X	X	X	
Power Valve (Loose, Sticking)					X		X		X	X	
Fuel-Inlet Needle & Seat (Loose, Leaking)	X	X	X	X	X	X				X	
Float (Leaking, Rubbing, Wrong Setting)	X	X	X	X	X	X		X		X	
Gaskets (Brittle, Improper Seal)	X	X			X	X		X			
Pump Discharge Port (Dirty, Plugged)						X	X		X		
Pump Diaphragm (Worn, Cut)						X	X		X		
Pump Ball Checks (Dirty, Sticking)						X	X		X		
Choke-Quality Adjustment (Vacuum Leak)									X		X
Choke Valve & Linkage (Dirty, Sticking, Damaged)									X		X
Secondary Linkage Adjustment						X	X		X	X	
Throttle Valves (Loose, Damaged, Sticking)	X	X				X		X			X

Chart deals with carburetor problems only and assumes engine is in good mechanical condition and in tune. Many ignition and carburetor problems can give the same symptoms so don't assume the fault lies with the carburetor.

11 REPAIR AND ADJUSTMENT

SO YOU THINK YOU'VE GOT A VACUUM LEAK?

Air leaks are commonly called *vacuum leaks;* however, this term is really inaccurate—vacuum can't really leak. Air leaks *into* the *vacuum*—intake manifold in our case—and that causes problems.

The first step in finding an air leak is recognizing it. Depending on the size of the leak, the symptoms are fairly simple. The engine will idle roughly or "hunt"—refuse to idle at one speed. You can check for an air leak by partially closing the choke; if the engine runs better, it has an air leak. If the leak is bad enough, the engine won't idle at all.

Roughness at idle is caused by the engine getting more air than it needs, leaning out the mixture. Vacuum (air) leaks show up more at idle than at higher engine speeds. This is simply because the engine is receiving so little fuel/air mixture that the added air drastically changes the fuel/air ratio. At engine speeds above idle the amount of air leaking in is such a small amount compared to the total fuel/air mixture that the leak seems to disappear.

Unlike oil leaks, air leaks can't be seen, so you've got to find some other way to locate them. Sometimes they can be heard, but it's really hard to pin point one without putting you ear near a hot, running engine—definitely *not* recommended. S how do you find one? Spra carb cleaner.

Keep in mind that carb cleane is extremely flammable an should only be used in a well ventilated area. It also has a ter rific appetite for paint—eve heat resistant—so keep th overspray to a minimum.

Use the kind of cleaner wit the long plastic tube that snap into the nozzle. This gives much more accurate spray. Wit the engine idling, spray th areas you think might be leak ing, one at a time. Just a shor spurt is all that is needed. B sure to let the air clear befor you spray again. If you spra near an air leak, the engine wi momentarily smooth out an gain rpm.

If you have no idea where th leak is, start with the most likel places. Check the vacuum-hos connections first. If you've re placed the carburetor, spra around the base gasket.

When checking for air leaks don't trust your eyes. A quic glance won't reveal a crack in vacuum hose or a split gasket. this test says you've got a lea at a vacuum connection, take careful look. If you still can't se anything after a close inspec

carburetor removal and installation. Flare-nut wrenches are a *necessity* if you are working with brass fuel-line nuts.

I can't overstress the importance of tagging and identifying each vacuum hose with a piece of tape or shipping tag. Take a few minutes to make a schematic diagram showing all the hookups. This will save a lot of grief, especially with complicated emission-control systems. Follow the kit instructions for removing and installing the carburetor.

Once you've removed the carburetor, use a holding fixture to prevent damage to the throttle plates. Many of these stands or fixtures are available, but 5/16-in. bolts and nuts work just fine on any carburetor with four mounting holes. To prevent wobble, use a second nut on each bolt to lock it to the throttle body.

1/Model 5210 on the left is used by GM; 5200 on the right is used by Ford. Carburetors are very similar in design. Most are used on 4-cylinder engines.

Read the Preparation for Carburetor Repair before starting. Mount the carburetor on a stand to protect the throttle plates. Use 5/16-in. nuts and bolts or plastic carburetor legs. It's a good idea to loosen the fuel-inlet fitting while the carburetor is still on the engine.

Remove the fuel-inlet fitting, gasket, filter and the spring. A new gasket and filter are in the repair kit. Some 5200s use a non-replaceable plastic filter.

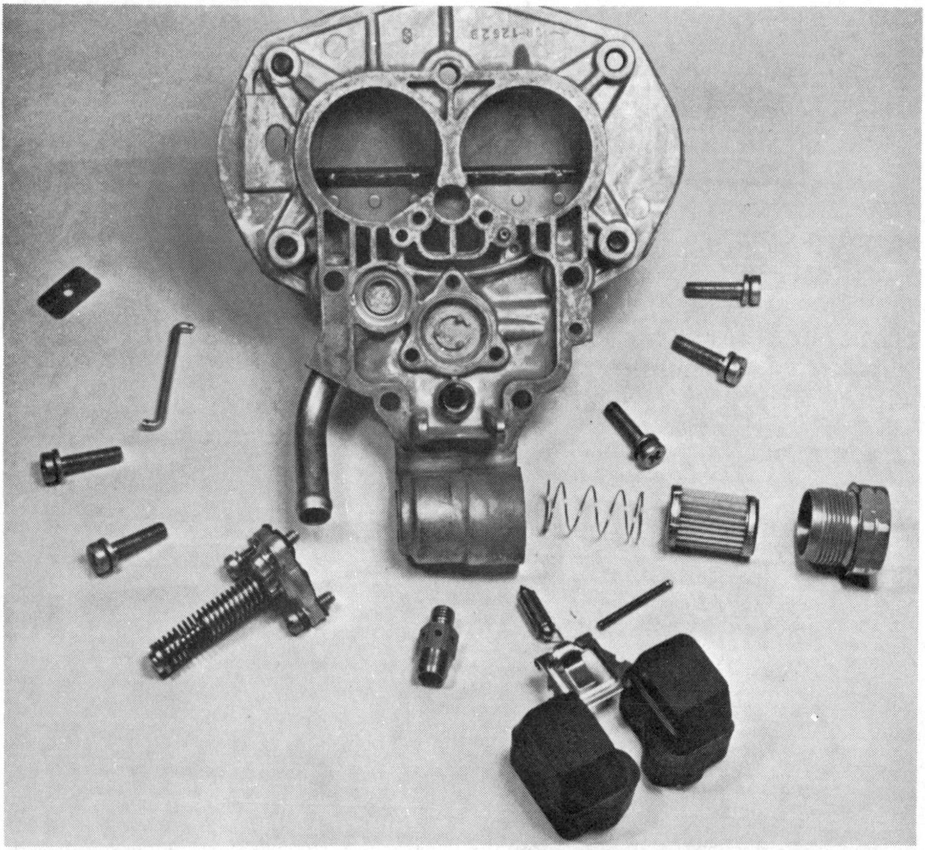

2/Remove the plastic retainers from both ends of the choke-operating link and free this link from both levers. Remove five screws holding the air horn to the main body and lift the air horn straight up. A slight rap from a plastic hammer or a screwdriver handle may be required to separate the two castings.

Slide the choke-operating link through the slot in the air horn, turn the casting to the side and slide the plastic guide out. Remove the float-hinge pin with a small punch. Lift out the float and inlet valve.

Remove the three screws holding the power valve and remove it. Now it will be easier to remove the fuel-inlet seat and gasket. Photo shows how far you usually have to disassemble the air horn.

If the choke plates or shaft are damaged or binding, remove them also. First, file the staking from the threaded ends and carefully remove the screws. Now slide the plates from the slotted shaft and the shaft from the air horn.

When you reinstall the choke plates, hold the air horn up to the light with the choke closed to make sure the plates are properly seated before tightening the attachment screws. Restake the screws with a prick punch or small chisel. Support the shaft while you do this to keep from bending the choke-plate shaft.

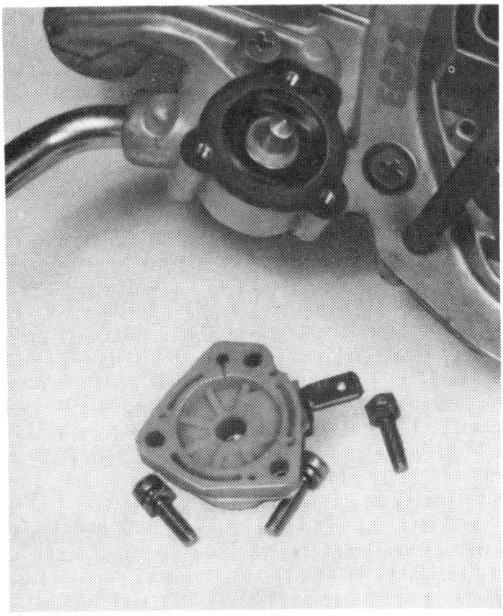

3/This bowl vent is vacuum controlled so it's closed while engine is running and open to the evaporative canister when engine is off. Shown alongside is a small solenoid that prevents the vent from opening under low-vacuum, heavy-load conditions. Don't immerse either the diaphragm or solenoid in harsh cleaner. These features are not on all 5200-series carburetors.

4/Before going further record the setting of the bimetal-housing pointer and the marks on the choke casting. You'll want to restore this setting on reassembly. As you face the housing, clockwise rotation gives a richer setting.

Index mark on the all-electric bimetal cap in white, and its relationship to the marks on the intermediate spacer.

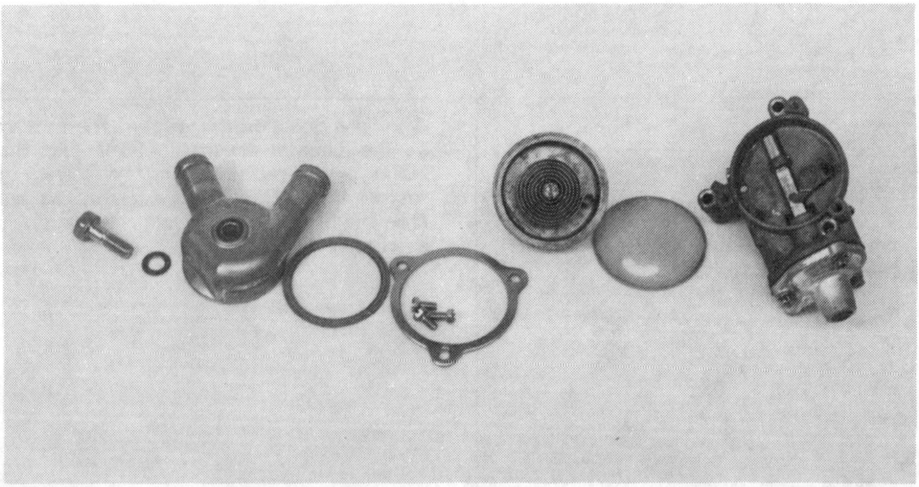

5/Remove the retaining bolt, washer, hot-water housing and gasket. This may require a rap from your screwdriver handle. Next, remove the three screws and retainer holding the bimetal housing and gasket.

Remove three screws holding the choke casting to main body. Disconnect the fast-idle link and remove the choke assembly.

Photo shows hot-water stove and bimetal assembly removed from the choke casting. Do not remove the bimetal from the housing or place the plastic gasket into harsh cleaner.

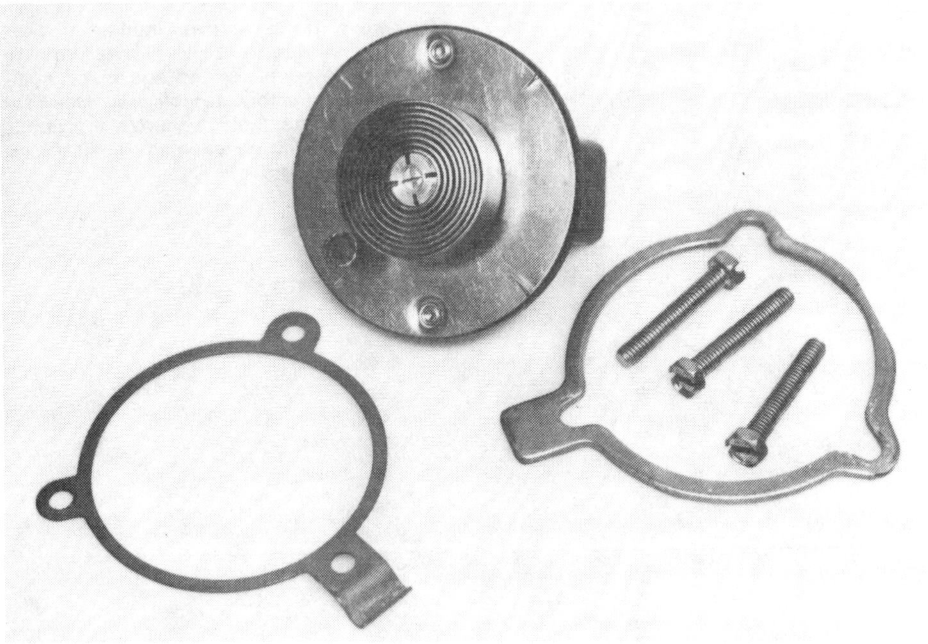

All-electric choke element is attached with three screws. Plastic body (not shown) is sandwiched between the two metal retainers. As with the hot-water choke, keep the body and the bimetal/heating unit out of any harsh cleaners or solvents.

6/Choke and diaphragm assembly. Replace O-ring (arrow), with a new one. This passage channels manifold vacuum to the choke diaphragm.

7/Remove three screws holding choke-diaphragm cover. This allows you to remove the cover, spring and diaphragm-and-stem assembly. Check diaphragm for holes or cracks. Unless there's a problem with the fast-idle mechanism this is as far as you have to go.

8/If you choose to go further, note the position of the spring between the fast-idle cam and the choke lever—check the photos. Remove the choke-shaft nut and lock washer. Now the choke lever, fast-idle cam, spring, cup, spacer, and the Teflon lever and the shaft can all be removed.

Remove the screw for the fast-idle lever, spacer and washer. Photo shows choke lever and fast-idle parts laid out in correct relation to one another.

9/Now for the main body of the carburetor. Remove the two screws holding the anti-dieseling-solenoid bracket (if used) to the main body.

Remove the main air bleeds and well tubes from the primary and secondary wells. Slip a bent paper clip into the tube, then pull it out. Ordinarily you should not put anything into a metering orifice but this one isn't critical. Record the number and position of each metering restriction as you pull it out.

10/Some carburetors have pressed-in idle-feed restrictions, others have screw-in jets. If yours has the screw-in type, remove primary and secondary idle jets and retainers on each side of the carburetor. Jet holder (retainer) contains the idle jet. Idle jets have a number stamped on one of the land surfaces.

11/Remove the main jets. Be sure to use a wide-blade screwdriver to avoid damaging the jets.

12/Remove the power valve. Do not bend the valve stem. Some carburetors use a three-piece power valve.

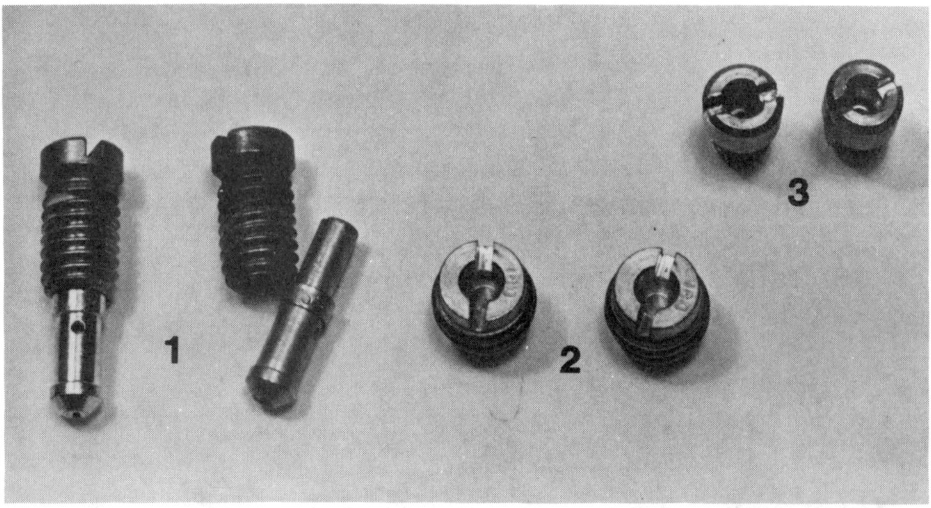

Record the number of each jet as you remove it and note its location so you can reinstall it in the same position. You may have to use a magnifying glass to read the numbers.

Some carburetors have idle-feed restrictions 1, with the primary one dyed red and the secondary one a natural brass color. Bleed restrictions 2, are stamped on top. Well tubes (not shown) are stamped on the bottom. Main jets 3, are stamped on the side.

13/Turn the limiter cap (if there is one) as far clockwise as it will go. Remove limiter cap with pliers, taking care not to bend the mixture screw.

Now turn the mixture screw in until it reaches its seat and record the number of turns. Use this as the idle setting for reassembly. Remove mixture screw and spring.

14/Remove pump-nozzle screw, nozzle, gaskets, and check balls. Some 5200s have two check balls. Remove four screws holding accelerator-pump housing to main body. Remove the housing, diaphragm and spring.

Normally, this is as far as you would go with the disassembly, especially with the economy method. If you insist on a complete job, be prepared to spend more time and follow the instructions closely.

On certain models of 5200-series carburetors, the secondary-throttle plate must be removed before the secondary throttle lever and spring. The lever is riveted to the shaft. Unless the throttle bore is quite dirty, the plates nicked or the shaft loose or sticky, don't do it. It is complicated and time-consuming and you risk stripping the threads in the throttle shaft.

Throttle-plate screws are staked on the threaded side and this staking must be removed with a file before taking out the screws. These screws must be re-staked with a small punch upon reassembly to prevent them from dropping into the engine. The throttle shaft must be supported during staking to prevent bending the shaft.

Before tightening the throttle-plate screws, hold the carburetor up to the light and make sure the plates are seated in the bore when both primary and secondary stop screws are backed off. As you're disassembling the throttle mechanism, pay close attention to how the parts relate to each other, particularly the return springs. The accompanying photographs will help here.

15/Bend the locking retainer tabs (arrows) away from the primary-throttle-shaft nut.

16/Remove retaining nut 1, and the entire mechanism attached to the primary throttle shaft can be removed. This includes the primary throttle lever 2, secondary operating lever 3, idle lever 4, return spring 5, and several spacers. Refer to exploded views, pages 94 and 95.

ASSEMBLY AND ADJUSTMENT

Now the carburetor is completely disassembled and ready for cleaning. Do the cleaning by either method, but remember: Only metal parts can be immersed in commercial-type carburetor cleaners.

Non-metal parts must be cleaned with a mild solvent such as kerosene or one of the other cleaners mentioned in Preparation for Carburetor Repair.

When you've thoroughly cleaned all of the components, blow out the passages with compressed air. If you don't have an air compressor, a hand tire pump will do nicely. Most metering restrictions are removable and easily cleaned. Never stick wires or drills into these restrictions.

Now inspect all of the components not included in the repair kit, and hence must be reused. Look for excessive wear. Pay particular attention to the two throttle shafts and the choke shaft. If the shafts are very loose, the complete carburetor will have to be replaced. This is very seldom the case.

Check the boost venturi for tightness. A loose fit here could also require carburetor replacement. For reassembly, work your way back through the disassembly procedure. The pictures will be of great assistance.

Several adjustments must be made during assembly, including dry float setting and rough settings for curb and fast idle. The choke system requires a number of adjustments, including choke index, dechoke, and vacuum quality, also known as vacuum "kick" or pulldown.

17/Be sure the return springs are positioned correctly before installing the carburetor, otherwise throttles may not close. These three photos should help, but before reinstalling the carburetor, check the throttles to be sure they close.

18/Choke-qualifying, fast-idle and choke-unloader adjustments all call for specific choke-plate-to-air-horn clearance. Measure clearance as shown using correct-diameter drill bits. Some car-manufacturers' specifications require checking the clearance at the upper edge of the choke plate.

19/Choke-qualify dimension. With a light closing force on the choke plate, push diaphragm stem toward diaphragm as far as it will go. With your third hand, measure choke-plate clearance between the lower edge of the plate and the casting using a drill (arrow). Correct dimension is listed in repair kit or Holley Spec Manual. Although the settings vary considerably, 0.25 in. (1/4 in.) is a good approximation.

20/Choke-qualify adjustment. Turning screw in gives a smaller clearance, or richer qualifying setting. Turning out has the opposite effect. Some 5200s have a slotted adjustment screw rather than the Allen type shown.

21/Dry-float setting is measured between the float lung and casting as shown—without gasket and with the air horn inverted.

Adjust by bending float tang contacting the inlet valve. Be careful not to mar the contact surface. Usually about 7/16 in., exact dimension is in the repair kit or Holley Spec Manual. Vehicle service manuals are another source.

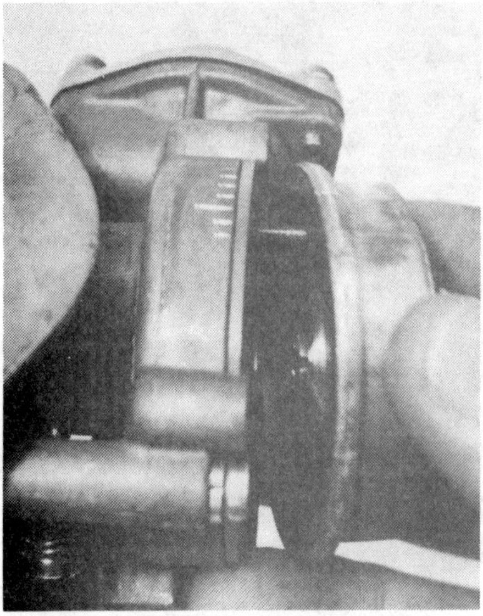

22/Make sure bimetal loop fits over the tang upon reassembly. Rotate bimetal housing to make sure choke plate moves.

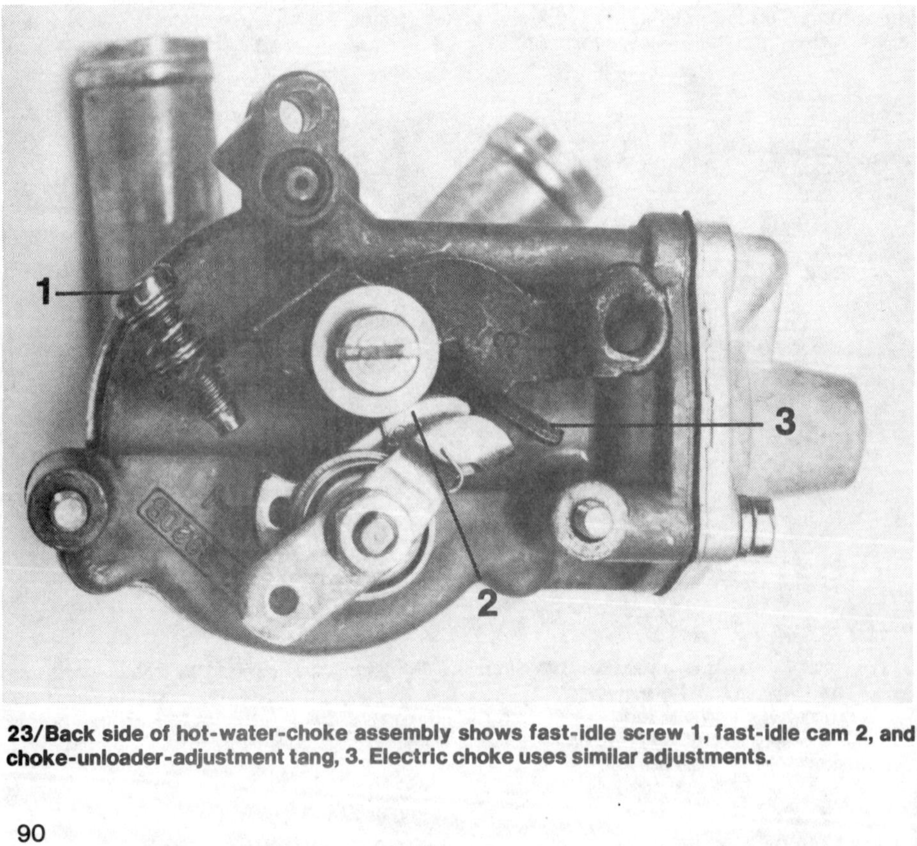

23/Back side of hot-water-choke assembly shows fast-idle screw 1, fast-idle cam 2, and choke-unloader-adjustment tang, 3. Electric choke uses similar adjustments.

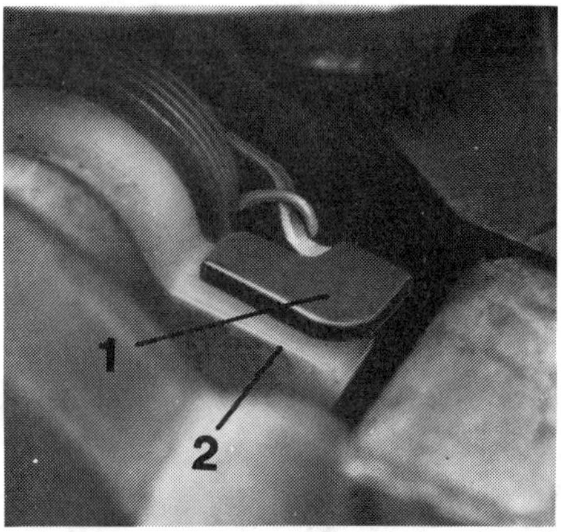

24/Fast-idle-cam phasing or index adjustment. Place a 5/32-in. drill between choke plate and casting as shown on page 88. With the fast-idle screw against the second-highest cam step, clearance between choke-lever tang 1, and the fast-idle cam 2, should be 0.010—0.030 in. Bend tang to adjust. This adjustment assures correct choke and fast-idle-cam relationship.

25/With the fast-idle adjustment screw against the top step of the cam, an initial fast-idle adjustment is 0.035-in. clearance between throttle plate and casting. Fast-idle screw is shown in step 23.

26/Measure the dechoke or unloader dimension between the choke plate and casting with the throttle held wide open. Dechoke settings vary considerably, but 0.25 in. (1/4 in.) is a good approximation. Consult repair manual or Holley Spec Manual for exact dimension. Adjustment tang is shown in step 23. On some 5200s the dechoke is set automatically when the fast-idle cam is indexed (step 24).

27/Check that the choke plates seat well in the housing and open and close without sticking or binding. With the air horn off, hold the choke plate and housing up to a light and check for uniform clearance.

28/Rough curb-idle adjustment 1 is 1-1/2 turns from where the plate is seated in the bore. You'll want to set to correct rpm on the engine. When an anti-dieseling solenoid is used, the idle speed is set with screw 2 with the solenoid activated.

29/Typical secondary idle-speed adjustment should be 1/4—1/2 turn from where the throttle plate is seated in the bore.

Model 5200 Models 5210 and 5220 are similar.

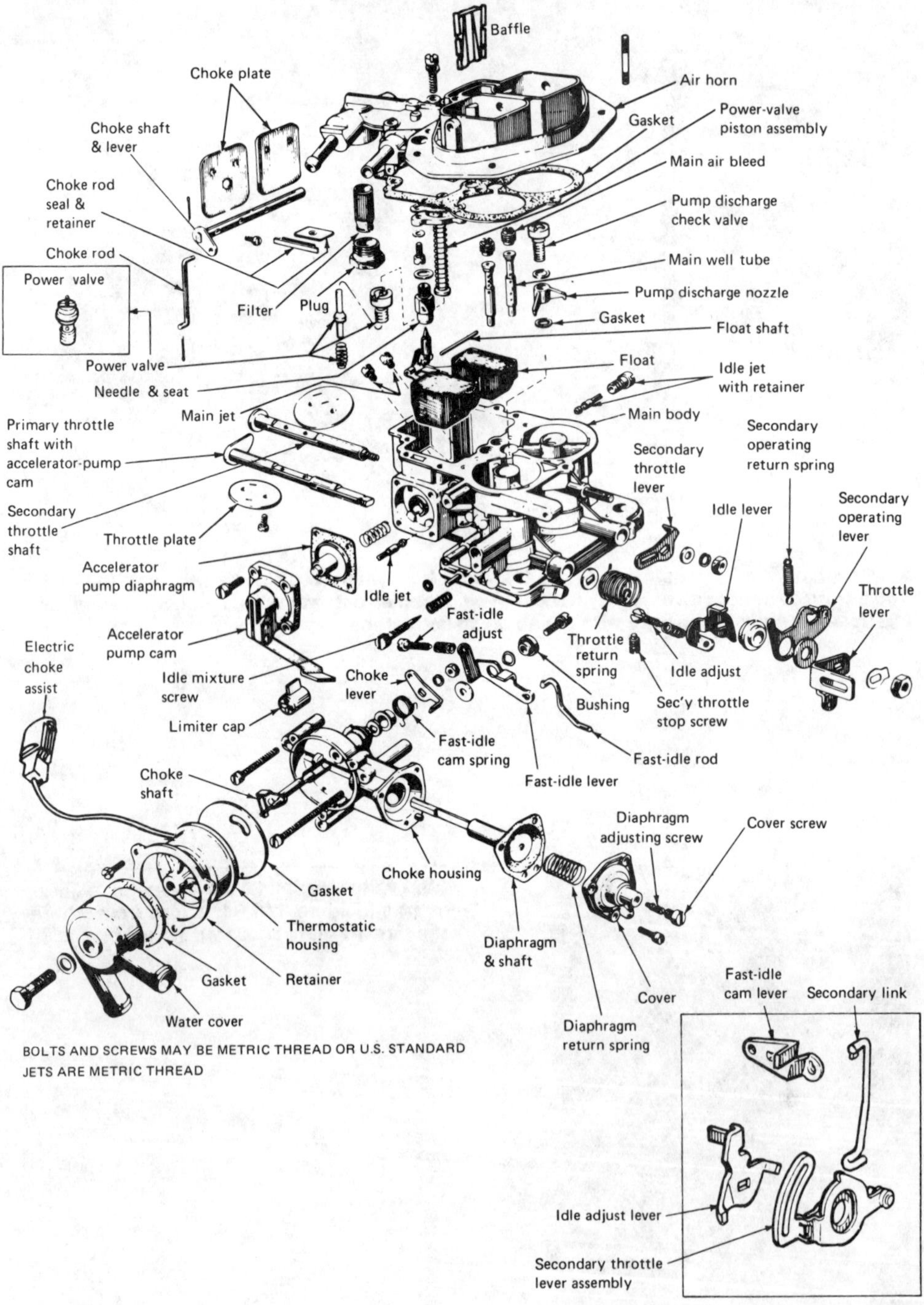

BOLTS AND SCREWS MAY BE METRIC THREAD OR U.S. STANDARD
JETS ARE METRIC THREAD

Model 6520 Models 6500 and 6510 are similar.

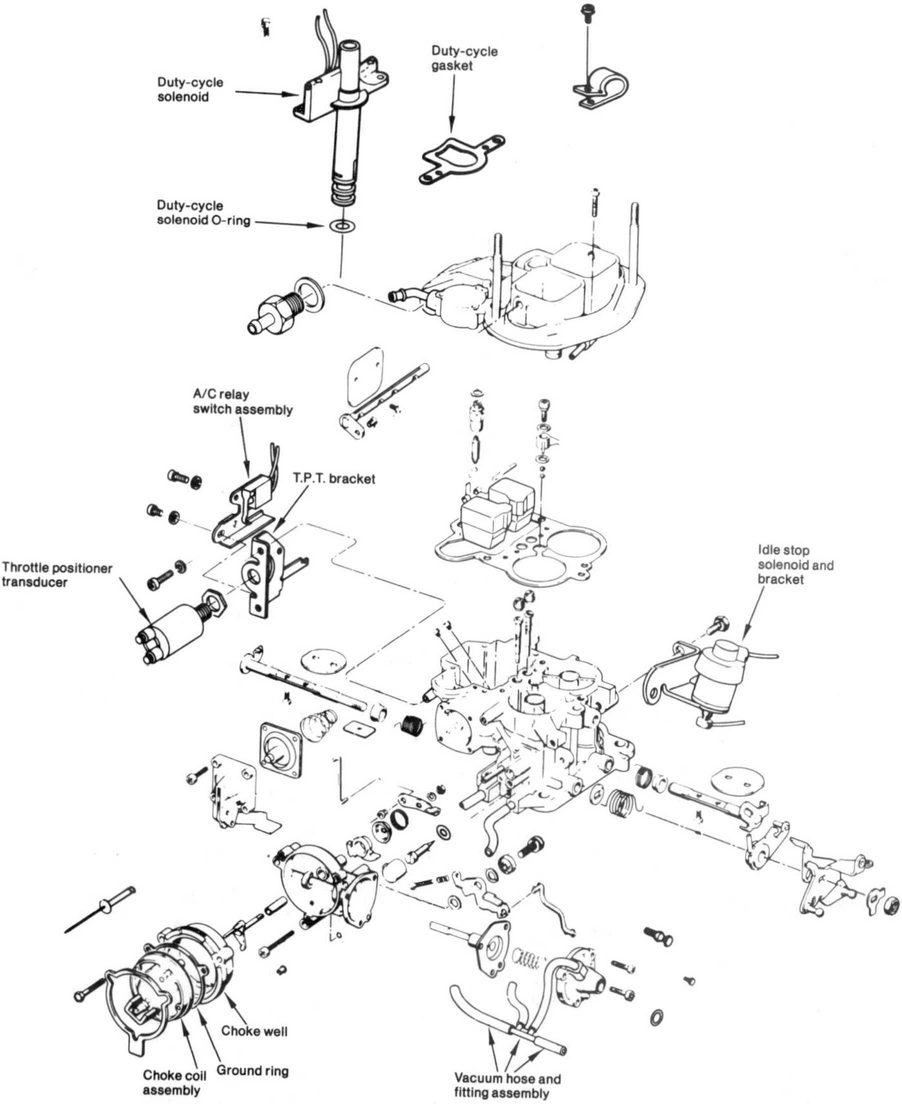

INDEX